Prealgebra Review Workbook

LAURA WHEEL

Boston   San Francisco   New York
London   Toronto   Sydney   Tokyo   Singapore   Madrid
Mexico City   Munich   Paris   Cape Town   Hong Kong   Montreal

> ⚠ This work is protected by United States copyright laws and is provided solely for the use of instructors in teaching their courses and assessing student learning. Dissemination or sale of any part of this work (including on the World Wide Web) will destroy the integrity of the work and is not permitted. The work and materials from it should never be made available to students except by instructors using the accompanying text in their classes. All recipients of this work are expected to abide by these restrictions and to honor the intended pedagogical purposes and the needs of other instructors who rely on these materials.

Reproduced by Pearson Addison-Wesley from electronic files supplied by the author.

Copyright © 2006 Pearson Education, Inc.
Publishing as Pearson Addison-Wesley, 75 Arlington Street, Boston, MA 02116.

All rights reserved. This manual may be reproduced for classroom use only.

ISBN   0-321-47332-9

4 5 6 BB 09 08

Table of Contents

Chapter 1   Integers
Section 1: Review of Whole Numbers.......................... 1
Section 2: Integers and Absolute Value...................... 17
Section 3: Operations with Integers......................... 21
Section 4: Exponents and Order of Operations................ 31
Section 5: Averages... 37

Chapter 2   Fractions
Section 1: Factors and Multiples 45
Section 2: Introduction to Fractions........................ 53
Section 3: Simplifying Fractions............................ 62
Section 4: Addition and Subtraction of Fractions............ 72
Section 5: Multiplication and Division of Fractions 88
Section 6: Exponents and Order of Operations................ 101
Section 7: Complex Fractions................................ 105

Chapter 3   Decimals
Section 1: Introduction to Decimals......................... 109
Section 2: Rounding and Comparing Decimals 116
Section 3: Addition and Subtraction of Decimals............. 121
Section 4: Multiplication and Division of Decimals.......... 128
Section 5: Fractions and Decimals........................... 141
Section 6: Scientific Notation.............................. 149
Section 7: Square Roots..................................... 160

Chapter 4   Basic Concepts of Algebra
Section 1: Variables and Algebraic Expressions 169
Section 2: Solving One-Step Equations....................... 176
Section 3: Solving Two-Step Equations 183
Section 4: Algebra and Problem Solving...................... 186

Chapter 5   Ratios, Rates, and Proportions
Section 1: Ratios... 191
Section 2: Rates.. 197
Section 3: Introduction to Proportions...................... 204
Section 4: Solving Proportions.............................. 208
Section 5: Problem Solving with Proportions................. 214

Chapter 6 Percents
    Section 1: Introduction to Percent................................... 221
    Section 2: Percents and Decimals.................................. 225
    Section 3: Percents and Fractions.................................. 228
    Section 4: Percent Equations.. 233
    Section 5: Problem Solving with Percents...................... 237
    Section 6: Mean, Median, and Mode............................. 249

Practice Tests
    Test A... 253
    Test B... 255
    Test C... 257

Answers to Exercises
    Chapters 1-6... A-1
    Practice Tests... A-7

Chapter 1   Integers

Section 1: Review of Whole Numbers

The **whole numbers** are 0, 1, 2, 3, 4, 5, 6, 7, 8, 9, 10, ...

The three dots mean that the numbers continue on indefinitely. There is no largest whole number and the smallest whole number is 0.

A digit is a number 0, 1, 2, 3, 4, 5, 6, 7, 8, or 9 that names a place-value location. The digits of large numbers are separated by commas into groups of three. Each group has a name: *ones, thousands, millions, billions, trillions,* and so on.

Place Value Chart												
Trillions			Billions			Millions			Thousands			Ones		
Hundreds	Tens	Ones	Hundreds	Tens	Ones	Hundreds	Tens	Ones	Hundreds	Tens	Ones	Hundreds	Tens	Ones

EXAMPLE 1   What digit is in the ten thousands place in the number 1,287,593?

SOLUTION   The digit in the ten thousands place is the fifth digit from the right, or 8.

The number 24,734 is written in **standard notation**. The number can be written in **expanded notation** as follows.

$$24{,}734 = 2 \text{ ten thousands} + 4 \text{ thousands} + 7 \text{ hundreds} + 3 \text{ tens} + 4 \text{ ones}$$
$$= 20{,}000 \quad + 4000 \quad + 700 \quad + 30 \quad + 4$$

The word name for a number is how the number is said.

EXAMPLE 2   Write the word name for 136,455,701.

SOLUTION   One hundred thirty-six million, four hundred fifty-five thousand, seven hundred one

CHAPTER 1 INTEGERS   PREALGEBRA REVIEW

Addition of Whole Numbers

The numbers that are added in an addition problem are called **addends** and the **sum** is the result of adding. Addition is denoted by the "+" symbol.

> **Procedure for Adding Whole Numbers:**
> To add two or more whole numbers, line up the numbers vertically by place value and then add the digits in each column starting with the ones column. When the sum in a column has two digits, carry the left digit of the sum to the top of the next column on the left.

EXAMPLE 3   Add: 795 + 121

SOLUTION

$$\begin{array}{r} \overset{1}{}795 \\ +\ 121 \\ \hline 916 \end{array}$$

Note that if the order of the addends is changed, the sum is not affected. In Example 3, changing the order of the addends does not affect the sum. That is, 795 + 121 = 916 and 121 + 795 = 916. This fact about addition is called the *commutative property of addition*.

Commutative Property of Addition: For any numbers a and b, $a + b = b + a$.

Note that the letters a and b, which are called **variables**, are used to represent numbers in the definition of the Commutative Property. Variables will be discussed later in the book.

EXAMPLE 4   Add: 1573 + 87 + 599

SOLUTION

$$\begin{array}{r} \overset{1\,2\,1}{1573} \\ 87 \\ +\ \ 599 \\ \hline 2259 \end{array}$$

When adding three or more numbers, the numbers can be grouped in any way without affecting the sum. In Example 4, 1573 and 87, can be added first, and then 599 can be added to that sum. That is, (1573 + 87) + 599 = 1160 + 599 = 2259. Or, 87 and 599 can be added first, and then 1573 can be added to that sum. That is, 1573 + (87 + 599) = 1573 + 686 = 2259. This fact about addition is called the *associative property of addition*. Parentheses are used to show the grouping of the addends.

Associative Property of Addition: For any numbers a, b, and c, $(a + b) + c = a + (b + c)$.

Subtraction of Whole Numbers

In a subtraction problem, the number from which another number is being subtracted is called the **minuend** and the number being subtracted is called the **subtrahend**. The **difference** is the result of subtracting. Subtraction is denoted by the "−" symbol.

Addition and subtraction are *inverse* operations because they undo each other. Consider $6 + 8 = 14$. To undo this addition, use subtraction as follows: $14 - 8 = 6$ or $14 - 6 = 8$.

> **Procedure for Subtracting Whole Numbers:**
> To subtract two whole numbers, line up the numbers vertically by place value with the larger number on top and then subtract the digits in each column starting with the ones column. Subtract the bottom digit from the top digit. If the top digit is smaller than the bottom digit, borrow from the top digit in the column to the left.

EXAMPLE 5 Subtract: $2832 - 1579$

SOLUTION

$$\begin{array}{r} {\scriptstyle 7\,12\,12} \\ 2\,8\,3\,2 \\ -1\,5\,7\,9 \\ \hline 1\,2\,5\,3 \end{array}$$

Note that the Commutative Property and Associative Property do apply to subtraction.

Multiplication of Whole Numbers

In a multiplication problem, the numbers that are multiplied are called **factors** and the **product** is the result of multiplying. Multiplication can be denoted by the "×" symbol, by a raised dot "·", or with parentheses. For instance, each of the following expressions represents four times five:

$$4 \times 5 \qquad 4 \cdot 5 \qquad (4)(5) \qquad 4(5)$$

Multiplication is a way of representing repeated addition. For instance, $4 \times 5 = \underbrace{5 + 5 + 5 + 5}_{\text{Add 5 four times}}$.

CHAPTER 1 INTEGERS   PREALGEBRA REVIEW

For the purposes of this discussion, it is assumed that the reader is familiar with multiplication of single digit numbers.

The product of a number and 0 is 0 and the product of a number and 1 is the number itself. These two facts about multiplication are summarized below.

Multiplication Property of 0: If a is any number, then $0 \cdot a = 0$ and $a \cdot 0 = 0$.

Multiplication Property of 1: If a is any number, then $1 \cdot a = a$ and $a \cdot 1 = a$.

Note that the number 1 is also called the *multiplicative identity*.

Multiplying a whole number by a single digit whole number:

$$\begin{array}{r} \overset{2}{2}8 \\ \times\ \ 3 \\ \hline 84 \end{array}$$

- Multiply 8 ones by 3: 3×8 (ones) = 24 ones = 2 tens and 4 ones. Put the 4 in the ones column and carry the 2 to the tens column.
- Multiply 2 tens by 3: 3×2 (tens) = 6 tens
  Add 2 tens: 6 tens + 2 tens = 8 tens
  Put the 8 in the tens column.

This process for multiplication is based on a property called the *distributive property*. The distributive property says how to multiply a number times a sum.

Distributive Property: For any numbers a, b, and c, $a(b+c) = a \cdot b + a \cdot c$.

Multiplying 28×3 using the distributive property gives

$$3 \cdot 28 = 3(20+8) = 3 \cdot 20 + 3 \cdot 8 = 60 + 24 = 84$$

To multiply two whole numbers, line up the numbers vertically by place value and then use the distributive property to multiply the top number by each digit of the bottom number.

PREALGEBRA REVIEW   CHAPTER 1 INTEGERS

EXAMPLE 6   Multiply: 3485×241

SOLUTION

$$\begin{array}{r} 3485 \\ \times\ 241 \\ \hline 3485 \\ 13940 \\ \underline{6970} \\ 839{,}885 \end{array}$$

Multiply 1×3485. Write the product leftward starting with the ones column.
Multiply 4×3485. Write the product leftward starting with the tens column.
Multiply 2×3485. Write the product leftward starting with the hundreds column.
Add.

If the order of the factors is changed, the product is not affected. In Example 6, changing the order of the factors does not affect the product. That is, $3485 \times 241 = 839{,}885$ and $241 \times 3485 = 839{,}885$. This fact about multiplication is called the *commutative property of multiplication*.

Commutative Property of Multiplication: If a and b are any number, then $a \cdot b = b \cdot a$

When multiplying three or more numbers, the numbers can be grouped in any way without affecting the product. For instance, in the product $4 \cdot 2 \cdot 3$, 4 and 2 can be multiplied first, and then 3 can be multiplied times that product. That is, $(4 \cdot 2) \cdot 3 = (8) \cdot 3 = 24$. Or, 2 and 3 can be multiplied first, and then 4 can be multiplied times that product. That is, $4 \cdot (2 \cdot 3) = 4 \cdot (6) = 24$. This fact about multiplication is called the *associative property of multiplication*. Parentheses are used to show the grouping of the factors.

Associative Property of Multiplication: For any numbers a, b, and c, $(a \cdot b) \cdot c = a \cdot (b \cdot c)$.

Division of Whole Numbers

The *quotient* is the result of dividing. Division can be denoted by the "$\div$" symbol, by the "$\overline{)}$" symbol, or by the "/" symbol. For instance, each of the following expressions represents 20 divided by 5:

$$20 \div 5 \qquad 5\overline{)20} \qquad 20/5 \qquad \frac{20}{5}$$

CHAPTER 1 INTEGERS   PREALGEBRA REVIEW

Multiplication and division are *inverse* operations because they undo each other. Consider $4 \times 5 = 20$. To undo this multiplication, use division as follows: $20 \div 5 = 4$ or $20 \div 4 = 5$. In the problem $20 \div 5 = 4$, 20 is called the **dividend**, 5 is called the **divisor**, and 4 is the **quotient.**

Division can be thought of in terms of repeated subtraction. For instance, to divide 20 by 5, repeatedly subtract 5 as follows:

$$20 - 5 = 15 \rightarrow 15 - 5 = 10 \rightarrow 10 - 5 = 5 \rightarrow 5 - 5 = 0$$

Notice that five is subtracted four times. Although division can be thought of in terms of repeated subtraction, it is not the fastest way to divide, especially when the numbers are large.

In some division problems, the divisor does not divide the dividend without leaving some amount left over. This left over amount is called a **remainder.** For instance, in the division $23 \div 5$, 5 divides into 23 four times with three leftover. That is, $23 \div 5 = 4$ R3. The letter R is used to indicate the remainder. Note that when the divisor is a factor of the dividend, the quotient has no remainder.

To divide large numbers, a process called *long division* is used.

$8\overline{)653}$
- Start with the digit in leftmost place in the dividend. Compare this digit with the divisor. If it is less than the divisor, then include the next digit, continuing until the digits in the dividend name a number that is greater than the divisor.

$\begin{array}{r} 8 \\ 8\overline{)653} \\ \underline{64} \\ 13 \end{array}$
- Divide 65 by 8: $65 \div 8 = 8$ (with a remainder of 1). Place the quotient, 8, above the tens digit in the dividend. Multiply: $8 \times 8 = 64$. Place 64 below 65 in the dividend. Subtract and write the next digit in the dividend, 3, next to the remainder 1. (This is commonly referred to as "bringing down" the 3).

$\begin{array}{r} 81 \\ 8\overline{)653} \\ \underline{64} \\ 13 \\ \underline{8} \\ 5 \end{array}$
- Divide 13 by 8: $13 \div 8 = 1$ (with a remainder of 5). Place the quotient, 1, above the ones digit in the dividend. Multiply: $8 \times 1 = 8$. Place 8 below 13. Subtract. There are no more digits in the dividend to bring down.

The quotient is 81 with a remainder of 5, which is written 81 R5. To check the result, multiply the quotient by the divisor, then add the remainder. That is, $81 \times 8 = 648$ and $648 + 5 = 653$.

Note that the Commutative Property and Associative Property do not apply to division.

PREALGEBRA REVIEW   CHAPTER 1 INTEGERS

EXAMPLE 7   Divide: $2317 \div 36$

SOLUTION

$$\begin{array}{r} 64 \\ 36{\overline{\smash{\big)}\,2317}} \\ \underline{216} \\ 157 \\ \underline{144} \\ 13 \end{array}$$

The quotient is 64 R13. Check: $64 \times 36 = 2304 + 13 = 2317$.

Some additional facts about division are as follows.

- Any number divided by 1 is the same number: $a \div 1 = \dfrac{a}{1} = a$

- Any number divided by itself is 1: $a \div a = \dfrac{a}{a} = 1$

- Zero divided by any nonzero number is 0: $0 \div a = \dfrac{0}{a} = 0, \quad a \neq 0$

- Division by 0 is not defined: $a \div 0$ or $\dfrac{a}{0}$ is not defined

Solving Applied Problems

Part of learning mathematics involves applying skills and techniques to solve applied problems. The strategy for solving such problems should include carefully reading the problem, understanding what is being asked, organizing the relevant given information, identifying the appropriate operation or operations needed to solve the problem, and writing expressions and equations. Some key words and phrases associated with the arithmetic operations that sometimes appear in applied problems are given in the chart.

Addition	Subtraction	Multiplication	Division
add	subtract	multiply	divide
sum	difference	product	quotient
plus	minus	times	divided by
more than	less than	multiplied by	
total	decreased by	of	
added to	subtracted from		
increased by			

CHAPTER 1   INTEGERS

EXAMPLE 8   The highest score on an exam was 97 and the lowest score was 56. What is the difference between the highest and lowest exam score?

SOLUTION   In the problem, the highest exam score and the lowest exam score are given. To find the *difference* between the highest and lowest exam score, subtract.

$$\begin{array}{rl} \text{Highest score} \rightarrow & 97 \\ -\text{ Lowest score} \rightarrow & -56 \\ \hline \text{Difference} & 41 \end{array}$$

The difference between the exam scores is 41 points.

EXAMPLE 9   A video rental store charges $4 for each movie rental? If Sam has $14, how many movies can he rent?

SOLUTION   The information given in the problem is the charge for each movie rental and the amount of money Sam has. Repeated subtraction of $4 for each rental will work (try it), but a faster way to arrive at the answer is to *divide* the amount of money Sam has by the cost of each movie rental.

Amount of money Sam has ÷ Cost of each movie rental   → $14 ÷ $4

$$14 \div 4 = 4\overline{)\begin{array}{r} 3 \\ 14 \\ \underline{12} \\ 2 \end{array}}$$

The quotient represents the number of movies Sam can rent. Note that the remainder represents the amount of money Sam has leftover, which is $2. So, Sam can rent 3 movies.

EXAMPLE 10 A sales representative drove 120 miles on Monday, 40 miles on Tuesday, 173 miles on Wednesday, and 89 miles on Friday. Find the total number of miles the sales representative drove over the four days.

SOLUTION In this problem, the number of miles the sales representative drove each of the four days is given. To find the *total* number of miles the sales representative drove, add.

$$\begin{array}{rl} \text{Monday} \rightarrow & \overset{2\,1}{1}20 \\ \text{Tuesday} \rightarrow & 40 \\ \text{Wednesday} \rightarrow & 173 \\ +\quad \text{Friday} \rightarrow & +\ 89 \\ \hline & 422 \end{array}$$

The sales representative drove a total of 422 miles over the four days.

EXAMPLE 11 Anita deposits $525 per month into her savings account. How much will she have in her account after 6 months?

SOLUTION The given information in this problem is the amount Anita deposits each month. The question asks how much she will have in her account after 6 months. Adding $525 six times will give the correct answer (try it), but a more efficient way to solve the problem is to *multiply* the amount Anita deposits by the number of months.

Amount Anita deposits each month × Number of months → $525 × 6

$$\begin{array}{r} \overset{1\,3}{5}25 \\ \times\ \ \ 6 \\ \hline 3150 \end{array}$$

After 6 months Anita has $3150 in her savings account.

CHAPTER 1   INTEGERS

EXAMPLE 12   A homeowner wants to tile her bathroom floor using 6 inch by 6 inch square tiles. If the floor is 6912 square inches, how many tiles will she need?

SOLUTION   The dimensions of the tiles and the area of the floor space are given. To find the number of tiles she will need, divide the area of the floor by the area of one tile. To find the area of one square tile, multiply.

Length of tile × Width of tile   →   6 inches × 6 inches

$6 \times 6 = 36$

So one tile has an area of 36 square inches. Now find the number of tiles she will need.

Area of floor ÷ Area of tile   →   6912 square inches ÷ 36 square inches

$$6912 \div 36 = 36 \overline{)\begin{array}{r} 192 \\ 6912 \\ \underline{36} \\ 331 \\ \underline{324} \\ 72 \\ \underline{72} \\ 0 \end{array}}$$

She will need 192 tiles.

EXAMPLE 13   Last year an employee's gross annual salary was $31,479. If a total of $8991 was withheld for taxes and social security, what was the employee's monthly take-home pay?

SOLUTION   The employee's annual gross salary and the amount withheld for taxes and social security are given. Deduct the amount withheld for taxes and social security from the employee's salary to find the employee's *take-home* salary last year. To do this, *subtract*.

Employee's gross annual salary   → $\overset{2\ 10\ 13\ 17}{\cancel{3}\cancel{1},\cancel{4}\cancel{7}9}$
−   Taxes and social security   →   − 8 9 9 1
                                    ─────────
                                    2 2, 4 8 8

PREALGEBRA REVIEW   CHAPTER 1   INTEGERS

The employee's take-home salary last year was $22,488. Note that there are 12 months in a year. To find the employee's monthly take-home pay, *divide*.

Employee's take-home salary ÷ 12   → $22,488 ÷ 12

$$22{,}488 \div 12 = 12\overline{)22{,}488}$$

```
           1 874
      12)22,488
         12
         10 4
          9 6
            8 8
            8 4
              4 8
              4 8
                0
```

The employee's monthly take-home pay was $1874.

CHAPTER 1 INTEGERS

Section 1: Exercises

Identify the digit in each place for the number **128,409,627**.

1. Hundred thousands
2. Millions
3. Ten millions
4. Hundreds

Write the number in expanded notation.

5. 12,307

6. 2,345,677

7. 4,031

8. 456,783,209

Write in standard notation.

9. 3 ten thousands + 8 thousands + 0 hundreds + 6 tens + 1 one

10. 5 millions + 9 hundred thousands + 9 ten thousands + 1 thousand + 4 hundreds + 2 tens + 6 ones

11. 8 hundreds + 5 tens + 0 ones

12. 6 hundred millions + 1 ten million + 1 million + 4 hundred thousands + 9 ten thousands + 4 thousands + 1 hundred + 1 ten + 6 ones

Write a word name.

13. 227,896

14. 256

15. 9,705

16. 13,508,743

PREALGEBRA REVIEW   CHAPTER 1   INTEGERS

Add.

17. $789 + 54$

18. $9278 + 589$

19. $1507 + 13{,}988$

20. $6684 + 4955$

21. $385 + 29 + 1187$

22. $1763 + 894 + 14{,}056$

Subtract.

23. $823 - 245$

24. $4906 - 3791$

25. $186{,}244 - 70{,}565$

26. $87{,}500 - 9881$

Multiply.

27. 374×2

28. 5×470

29. $821(56)$

30. $39(1202)$

31. $512 \cdot 294$

32. $(2078)(453)$

Divide.

33. $674 \div 4$

34. $8571 \div 9$

35. $\dfrac{975}{15}$

36. $48\overline{)12{,}688}$

37. $328\overline{)57{,}584}$

38. $805{,}629 \div 251$

Choose the correct answer.

39. What is the remainder when 7943 is divided by 26?

    (A) 0   **(B)** 13   **(C)** 26   **(D)** 305   **(E)** None of these

CHAPTER 1   INTEGERS   PREALGEBRA REVIEW

40. When 512 is divided by a number, the quotient is 32. What is the divisor?

    (A) 16   **(B)** 32   **(C)** 0   **(D)** 512   **(E)** None of these

41. Which fact about numbers is demonstrated in the expression $4(3 + 8) = 4(8 + 3)$?

    (A) Commutative property of addition
    (B) Associative property of addition
    (C) Commutative property of multiplication
    (D) Associative property of multiplication
    (E) Distributive property

42. Which fact about numbers is demonstrated in the expression $4(3+8) = 4\cdot 3 + 4\cdot 8$?

    (A) Commutative property of addition
    (B) Associative property of addition
    (C) Commutative property of multiplication
    (D) Associative property of multiplication
    (E) Distributive property

43. What is the product of 420 and 4?

    (A) 424   **(B)** 416   **(C)** 1680   **(D)** 105   **(E)** 1660

44. What is the difference between 64 and 16?

    (A) 80   **(B)** 48   **(C)** 1024   **(D)** 4   **(E)** 58

45. How many weeks are in 91 days?

    (A) 637   **(B)** 11   **(C)** 18   **(D)** 7   **(E)** 13

46. How many days are in 4 weeks?

    (A) 20   **(B)** 14   **(C)** 28   **(D)** 7   **(E)** 4

47. How many minutes are in 3 hours?

    (A) 180   **(B)** 360   **(C)** 90   **(D)** 20   **(E)** 300

48. How many yards are in 102 feet?

    (A) 10   **(B)** 32   **(C)** 306   **(D)** 34   **(E)** 51

PREALGEBRA REVIEW

CHAPTER 1   INTEGERS

49. A customer receives a bill for service from a car repair shop. The total charge for 3 hours of labor was $135. How much did the repair shop charge for each hour of labor?

   (A) $405   **(B)** $45   **(C)** $68   **(D)** $40   **(E)** $138

50. An employee was paid $11,980 during the first half of the year. During the second half, she was paid $14,921. How much more was her income during the second half of the year?

   (A) $2941   **(B)** $26,601   **(C)** $3041   **(D)** $2959   **(E)** $2041

51. Elena had $782 in her checking account. She deposited a $397 pay check and a $529 pay check. What was the new balance in her checking account? (Assume no withdrawals were made.)

   (A) $1708   **(B)** $926   **(C)** $1698   **(D)** $1608   **(E)** $1311

52. How many square feet of carpeting are needed for a room that measures 11 feet by 12 feet?

   (A) 23   **(B)** 144   **(C)** 132   **(D)** 121   **(E)** 49

53. After graduating, Wes plans to drive from New York City to Los Angeles, a distance of 2758 miles, in one week. If he drives the same number of miles each day, what distance will he travel in one day?

   (A) 550   **(B)** 346   **(C)** 275   **(D)** 294   **(E)** 394

54. On a vacation trip, a family drove from Albany, NY to Orlando, FL, a distance of 1200 miles. If they drove 400 miles each day at a constant speed of 50 miles per hour, how many hours did they drive each day?

   (A) 3   **(B)** 8   **(C)** 24   **(D)** 6   **(E)** 12

55. A landscaper bought 4 rhododendron plants at $4 each, 6 hydrangeas at $7 each, and 3 lilac bushes at $9 each. Find the total cost of the plants.

   (A) $600   **(B)** $33   **(C)** $85   **(D)** $77   **(E)** $91

56. A contractor must tile an 1800 square inch countertop using 4 inch by 4 inch tiles. How many tiles will the contractor need?

   (A) 450   **(B)** 16   **(C)** 112   **(D)** 113   **(E)** 225

57. Kris bought a $1295 television set using a layaway plan. She put $275 down and paid the rest in equal monthly payments. If she paid each month for one year, how much was each payment?

   (A) $85   **(B)** $1020   **(C)** $130   **(D)** $100   **(E)** $88

58. A DVD movie is 117 minutes long. If the movie starts at 7:24 p.m., at what time will it end?

   (A) 8:30 p.m.   **(B)** 8:57 p.m.   **(C)** 9:21 p.m.   **(D)** 9:07 p.m.   **(E)** 8:41 p.m.

PREALGEBRA   CHAPTER 1 INTEGERS

Section 2: Integers and Absolute Value

The **integers** are …–5, –4, –3, –2, –1, 0, 1, 2, 3, 4, 5, …

The three dots on either side indicate that the integers continue on indefinitely in both directions. Numbers to the right of 0 are *positive numbers* and numbers to the left of 0 are *negative numbers*. The number 0 is neither positive or negative. A negative number is denoted by "–" symbol. A positive number is denoted by the "+" symbol. Note that a positive number can be written with or without the "+" symbol.

Many real-life quantities can be represented using positive and negative numbers. Some examples are as follows.

- A loss of $600 in the stock market can be represented as –$600.
- A hot air balloon flies at an altitude of 2500 feet above the ground. This altitude can be represented as +2500 feet, or 2500 feet.
- The temperature dropped 15° in the last hour. The drop in temperature can be represented as –15°.

Comparing Integers

The integers can be represented visually on a **number line.**

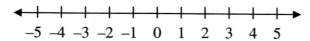

A number line can be used to compare numbers. On the number line, numbers are written in order, increasing from left to right. For any two numbers on the number line, the number to the left is *less than* the number to the right. Likewise, the number to the right is the *greater than* the number to the left. The inequality symbol > means "is greater than" and the inequality symbol < means "is less than." For instance, the statement –2 > –4 is read "–2 is greater than –4."

EXAMPLE 1   Write > or < between each pair of numbers to make a true statement.

    (a) 3 _____ –1   **(b)** –9 _____ –7   **(c)** 0 _____ 6

SOLUTION

    (a) 3 > –1   **(b)** –9 < –7   **(c)** 0 < 6

17

CHAPTER 1   INTEGERS

EXAMPLE 2   List the numbers 12, –3, –5, 8, 1, –10, 0 in order from smallest to largest.

SOLUTION   –10, –5, –3, 0, 1, 8, 12

Absolute Value

The **absolute value** of a number is its distance from 0 on the number line. The absolute value of a number is denoted by $|a|$, where a is any number. For instance, $|-1|$ is read as "the absolute value of –1." Since absolute value indicates a distance, the absolute value of a number is always positive or 0.

EXAMPLE 3   Determine the absolute value of the number.

   (a) $|9|$ (b) $|-10|$ (c) $|0|$ (d) $|-4|$

SOLUTION

   (a) 9   (b) 10   (c) 0   (d) 4

Opposites

Numbers that are the same distance from 0 but that are on opposites sides of 0 on the number line are called **opposites.** For example, on the number line, 2 lies two units to the right of zero and –2 lies two units to the left of zero. So, 2 and –2 are opposites.

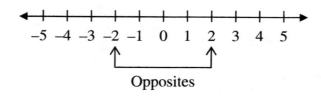

The sum of a number and its opposite is always 0. Because of this, the opposite of a number is also called its *additive inverse*. Note that 0 is its own additive inverse.

EXAMPLE 4   Find the opposite of the number.

   (a) –8   (b) 21

SOLUTION

   (a) 8   (b) –21

PREALGEBRA REVIEW CHAPTER 1 INTEGERS

The opposite of a number can also be found by placing a negative sign in front of the number. That is, the opposite of 6 is –6 and the opposite of –7 is –(–7), or 7. In general, the opposite of a number a is denoted as $-a$.

EXAMPLE 5 Simplify.

   (a) –(–4)   (b) –(–(–1))   (c) –|–9|

SOLUTION

   (a) –(–4) = 4
   (b) –(–(–1)) = –(1) = –1
   (c) Since |–9| = 9, –|–9| = –9

CHAPTER 1 INTEGERS

PREALGEBRA REVIEW

Section 2: Exercises

Express each quantity as a positive or negative number.

1. Mt. McKinley, the highest elevation in North America, is 6194 meters above sea level.

2. John's first round golf score was 4 below par.

3. Tamara lost 43 pounds through diet and exercise.

4. The N.Y. Giants gained 8 yards in the first play of the game.

Write > or < between each pair of integers to make a true statement.

5. 13 _____ –13   6. –4 _____ –3   7. –6 _____ 0   8. –15 _____ –17

List the numbers in order from smallest to largest.

9. 7, –8, 2, 11, –19, –9, 6, –1   10. –4, –7, 0, 16, –14, 9, –11, 4

Find the absolute value of the number.

11. $|-8|$ 12. $|5|$ 13. $|2|$ 14. $|-17|$

Insert >, <, or = to make a true statement.

15. $|-1|$ _____ –1   16. $|7|$ _____ $|-7|$ 17. $|-8|$ _____ $|-10|$ 18. –3 _____ $|-5|$

Determine the opposite of the number.

19. 11   20. –9   21. –14   22. 0

Simplify.

23. –(–10)   24. –(–(–8))   25. –$|5|$ 26. –$|-13|$

Section 3: Operations with Integers

Addition of Integers

The number line can be used to illustrate addition of integers. To add $a + b$ using the number line, start at a and move b units

- to the right if b is positive,
- to the left if b is negative.

If b is 0, then stay at a. The sum is the number reached after moving b units.

EXAMPLE 1   Add.

(a) $3 + (-5)$ (b) $-1 + (-6)$ (c) $-4 + 7$

SOLUTION

(a) $3 + (-5) = -2$

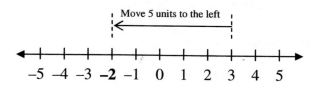

(b) $-1 + (-6) = -7$

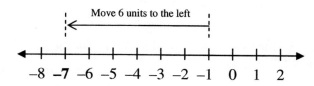

(c) $-4 + 7 = 3$

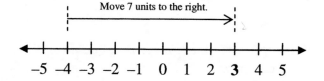

CHAPTER 1 INTEGERS

> **Procedure for Adding Integers:**
> - *Addends have the same sign* – Find the absolute values of the numbers and add them. Use the common sign as the sign of the sum. If both addends are positive, the sum is positive. If both addends are negative, the sum is negative.
> - *Addends have different signs* – Find the absolute values of the numbers and subtract the smaller absolute value from the larger absolute value. Use the sign of the number with the larger absolute value as the sign of the sum.

Since it is not always convenient to use a number line to add integers, the following procedure is used to add integers.

EXAMPLE 2 Add.

(a) $-58 + 35$ (b) $21 + (-17)$ (c) $-13 + (-12)$

SOLUTION

(a) Find the absolute values of the numbers.

$|-58| = 58$ and $|35| = 35$

Subtract 35 from 58: $58 - 35 = 23$

Since -58 has the larger absolute value and is negative, the sum is negative: $-58 + 35 = -23$

(b) $|21| = 21$ and $|-17| = 17$

Subtract: $21 - 17 = 4$

Since 21 has the larger absolute value and is positive, the sum is positive: $21 + (-17) = 4$

(c) $|-13| = 13$ and $|-12| = 12$

Add: $13 + 12 = 25$

Since both addends are negative, the sum is negative: $-13 + -12 = -25$

Recall from Section 2 that the opposite of a number is called its additive inverse because adding any number to its additive inverse always results in the sum 0. For instance, $3 + (-3) = 0$ and $-5 + 5 = 0$. In general, for any number a, $a + (-a) = 0$ or $-a + a = 0$.

EXAMPLE 3 Add: $8 + (-9) + 16 + (-11)$

SOLUTION By the commutative and associative properties of addition, the positive numbers can be grouped together and the negative numbers can be grouped together and then added separately.

$$8 + (-9) + 16 + (-11) = 8 + 16 + (-9) + (-11)$$
$$= 24 \;+\; -20$$
$$= 4$$

Subtraction of Integers

Consider the following subtraction and addition problems.

$6 - 4 = 2$ $6 + (-4) = 2$
Subtract 4   Add the opposite of 4

$9 - 1 = 8$ $9 + (-1) = 8$
Subtract 1   Add the opposite of 1

$7 - 3 = 4$ $7 + (-3) = 4$
Subtract 3   Add the opposite of 3

Each subtraction problem has the same answer as a related addition problem. In general, subtraction can be achieved by adding the opposite, or additive inverse, of the number being subtracted.

Procedure for Subtracting Two Integers:
To subtract $a - b$, add the opposite, or additive inverse, of b to a. That is, $a - b = a + (-b)$.

EXAMPLE 4 Subtract.

    (a) $18 - 25$ **(b)** $-7 - 10$ **(c)** $2 - (-14)$

SOLUTION

    (a) $18 - 25 = 18 + (-25) = -7$
    (b) $-7 - 10 = -7 + (-10) = -17$
    (c) $2 - (-14) = 2 + 14 = 16$

Multiplication of Integers

The following pattern can be used to demonstrate the product of a negative number and a positive number.

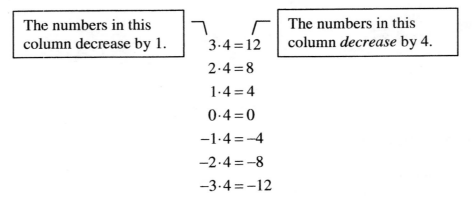

From the pattern, it appears that the product of a negative number times a positive number is negative. Likewise, the product of a positive number times a negative number is negative.

The following pattern can be used to demonstrate the product of two negative numbers.

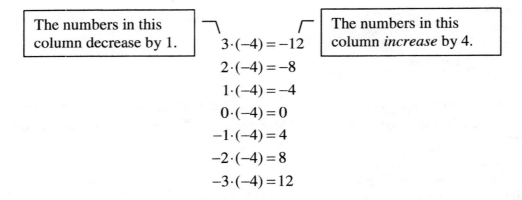

From the pattern, it appears that the product of a negative number times a negative number is positive.

> **Sign Rules for Multiplying Two Integers:**
> - *Same sign* – If the two numbers have the same sign, then the product is positive.
> - *Different signs* – If the two numbers have different signs, then the product is negative.

Recall from Section 1 that the multiplication property of 0 states that the product of 0 and a number is 0. That is, for any number a, $a \cdot 0 = 0$ and $0 \cdot a = 0$.

PREALGEBRA REVIEW   CHAPTER 1 INTEGERS

EXAMPLE 5   Multiply.

    (a) $7(-6)$ **(b)** $-1 \cdot 10$ **(c)** $-8 \cdot (-5)$ **(d)** $-13 \cdot 0$

SOLUTION

    (a) $7(-6) = -42$
    (b) $-1 \cdot 10 = -10$
    (c) $-8 \cdot (-5) = 40$
    (d) $-13 \cdot 0 = 0$

Note that multiplying a number times -1 as in Example 5(c) results in the opposite of the number. In general, for any number a, $-1 \cdot a = -a$.

EXAMPLE 6   Multiply: $-2 \cdot (-7) \cdot 4 \cdot (-5)$

SOLUTION

$$-2 \cdot (-7) \cdot 4 \cdot (-5) = 14 \cdot 4 \cdot (-5)$$
$$= 56 \cdot (-5)$$
$$= -280$$

Division of Integers

Recall that multiplication and division are inverse operations because they undo each other. This relationship can be used to develop the sign rules for division. Consider the following examples.

$$-3 \cdot 4 = -12, \text{ therefore } -12 \div 4 = -3$$
$$3 \cdot (-4) = -12, \text{ therefore } -12 \div (-4) = 3$$
$$-3 \cdot (-4) = 12, \text{ therefore } 12 \div (-4) = -3$$

From the examples, it appears that the quotient of a negative number divided by a positive number is negative. Similarly, the quotient of a positive number divided by a negative number is negative. And, the quotient of a negative number divided by a negative number is positive.

Sign Rules for Dividing Two Integers:
- *Same sign* – If the two numbers have the same sign, then the quotient is positive.
- *Different signs* – If the two numbers have different signs, then the quotient is negative.

CHAPTER 1 INTEGERS   PREALGEBRA REVIEW

EXAMPLE 7   Divide.

    (a) $36 \div (-2)$ **(b)** $-63 \div 7$ **(c)** $-175 \div (-25)$ **(d)** $0 \div (-11)$

SOLUTION

    (a) $36 \div (-2) = -18$
    (b) $-63 \div 7 = -9$
    (c) $-175 \div (-25) = 7$
    (d) $0 \div (-11) = 0$

Solving Applied Problems

EXAMPLE 8   Noi's checking account shows a current balance of –$47. What will be her new balance if she deposits $75?

SOLUTION   Noi's current balance and amount she deposits are given. Note that a deposit represents adding money to the account. To find the new balance, add the deposit amount to the current balance.

Current balance + Deposit amount →   –$47 + $75

$-47 + 75 = 28$

The new balance in Noi's checking account is $28.

EXAMPLE 9   The average surface temperature of the planet Jupiter is –166° Fahrenheit. The average surface temperature of the planet Mars is –65° Fahrenheit. What is the difference between these temperatures?

SOLUTION   The average surface temperatures for Jupiter and Mars are given. To find the difference between the temperatures, subtract.

Surface temperature − Surface temperature   → $-166° - (-65°)$
    of Jupiter   of Mars

$-166 - (-65) = -101$

The difference between the temperatures is –101° Fahrenheit. Note that this means that the average surface temperature of Jupiter is 101° colder than the average surface temperature of Mars.

PREALGEBRA REVIEW

CHAPTER 1 INTEGERS

EXAMPLE 10 A homeowner has his monthly mortgage payment of $827 automatically deducted from his checking account. Express the total amount deducted from his checking account in one year as a negative integer.

SOLUTION The amount of the monthly mortgage payment is given. The monthly mortgage payment can be expressed as a negative number, since it is deducted from his account. To find the total amount deducted from his checking account in one year, or 12 months, multiply.

Number of months · Monthly mortgage payment → 12 · (–$827)

12 · (–827) = –9924

The amount deducted from his checking account in one year can be expressed as –9924. This means a total of $9924 was deducted from his checking account in one year.

EXAMPLE 11 A woman weighing 244 pounds joined a weight loss program. After one year, she reached her goal weight of 140 pounds. Express her weekly weight change in pounds as a negative integer.

SOLUTION Both the woman's starting weight and goal weight are given. To find her weekly weight loss in one year, or 52 weeks, first find the total number of pounds she lost.

Starting weight – Goal weight → 244 pounds – 140 pounds

244 – 140 = 104

She lost a total of 104 pounds. That is her weight change was –104 pounds in 52 weeks. Now to find the weekly weight change, divide.

Weight change in one year ÷ Number of weeks → –104 pounds ÷ 52

–104 ÷ 52 = –2

Her weekly weight change in pounds can be expressed as –2 pounds. This means that she lost 2 pounds per week on the weight loss program.

CHAPTER 1   INTEGERS   PREALGEBRA REVIEW

Section 3: Exercises

Perform the indicated operation.

1. $15 + (-9)$ **2.** $-7 + (-12)$ **3.** $23 - 32$ **4.** $17 - 18$

5. $-56 + (-16)$ **6.** $-89 + 54$ **7.** $3 - (-15)$ **8.** $-39 - (-39)$

9. $-89 - (-98)$ **10.** $46 - (-73)$ **11.** $-106 + (-67)$ **12.** $-91 + 91$

Add.

13. $-23 + 17 + (-5)$ **14.** $31 + (-45) + 13$

15. $15 + (-8) + 1 + (-19)$ **16.** $-11 + 0 + (-3) + 21$

Perform the indicated operation.

17. $3(-9)$ **18.** $-6(12)$ **19.** $-56 \div 14$ **20.** $132 \div (-11)$

21. $-176 \div (-8)$ **22.** $-65 \div (-5)$ **23.** $(-9)(-15)$ **24.** $(-22)(-4)$

25. $-43(0)$ **26.** $-67(-1)$ **27.** $48 \div (-16)$ **28.** $-84 \div 7$

Multiply.

29. $(6)(-3)(2)$ **30.** $(-7)(4)(-1)$

31. $-5 \cdot (-8) \cdot 10 \cdot (-3)$ **32.** $2 \cdot (-11) \cdot (-2) \cdot 9$

Choose the correct answer.

33. What is the sum of -45 and 63?

   (A) -28 (B) 18 (C) -18 (D) 108 (E) -108

34. Find the sum of -16 and its opposite.

   (A) 32 (B) 16 (C) 0 (D) -32 (E) -16

PREALGEBRA REVIEW   CHAPTER 1   INTEGERS

35. Subtract –8 from its opposite.

    (A) 16   **(B)** 0   **(C)** –16   **(D)** 8   **(E)** –8

36. Divide 108 by –12.

    (A) 9   **(B)** 96   **(C)** 120   **(D)** –9   **(E)** –96

37. If the distance between two numbers on the number line is 5 units and one of the numbers is –2, then what is the other number?

    (A) 7 only   **(B)** 3 or –7   **(C)** 7 or –7   **(D)** 3 only   **(E)** –3 or –7

38. If two numbers are the same number of units from 3 on the number line and one of the numbers is –8, then what is the other number?

    (A) 11   **(B)** 8   **(C)** 14   **(D)** –11   **(E)** –2

39. What number when added to 8 results in a sum of –6?

    (A) 2   **(B)** 14   **(C)** –12   **(D)** –14   **(E)** –8

40. If the sum of –15 and 7 is increased by 12, what is the result?

    (A) –8   **(B)** 4   **(C)** –10   **(D)** –20   **(E)** 20

41. At noon the temperature was 19°. By 8:00 P.M. the temperature had dropped 24°. What was the temperature at 8:00 P.M.?

    (A) 43°   **(B)** 5°   **(C)** –5°   **(D)** –13°   **(E)** –6°

42. On first down, a football team lost 7 yards. On second down, the team lost 2 yards. On third down, the team gained 8 yards. What was the net gain in yards on the first three downs?

    (A) –1 yard   **(B)** –9 yards   **(C)** 1 yard   **(D)** 17 yards   **(E)** 8 yards

43. Anita has a balance $157 in her checking account. What is her new balance after she deposits $325 and then writes a $550 check for rent?

    (A) $482   **(B)** –$68   **(C)** $1032   **(D)** –$78   **(E)** $32

CHAPTER 1 INTEGERS   PREALGEBRA REVIEW

44. Sid's company deducts $83 from each paycheck for his health plan. If Sid is paid every two weeks, how much is deducted for his health plan in one year?

   (A) $1992   **(B)** $2158   **(C)** $4316   **(D)** $996   **(E)** $1079

45. An instructor deducts four points for each incorrect answer on an exam. A student has six incorrect answers. Express the number of points deducted on his exam as a negative integer.

   (A) –10   **(B)** –6   **(C)** –24   **(D)** –18   **(E)** –4

46. A store owner lost $47,264 in sales over the eight weeks her store was closed for renovations. If the store is usually open every day of the week and if she lost the same amount in sales each day, what was her daily loss expressed as a negative integer?

   (A) –6752   **(B)** –844   **(C)** –5908   **(D)** –1182   **(E)** –854

47. A meteorologist noted that the difference between the high and low temperatures on Tuesday was 37°. If the low temperature was –18°, what was the high temperature that day?

   (A) –19°   **(B)** 55°   **(C)** 19°   **(D)** –55°   **(E)** 45°

48. The air temperature drops about 3° for every thousand feet of change in altitude. If the temperature on the ground is 35°, what will the air temperature be at an altitude of 25,000 feet?

   (A) –40°   **(B)** 10°   **(C)** 40°   **(D)** –110°   **(E)** 5°

49. A hardware manufacturer gives its customers a $10 discount for every hundred dollars spent. A case of bolts costs $20. If a customer purchases 55 cases, how much will he pay after the discount is applied?

   (A) $1100   **(B)** $200   **(C)** $550   **(D)** $110   **(E)** $990

50. Over a five week period, the weight of a dieter fluctuated as follows: –1 pound, –2 pounds, +3 pounds, +1 pound, –4 pounds. What was his overall change in weight after five weeks?

   (A) 3   **(B)** 0   **(C)** –2   **(D)** 11   **(E)** –3

Section 4: Exponents and Order of Operations

Exponents

Expressions that contain repeated multiplication of the same factor, such as $3 \cdot 3 \cdot 3 \cdot 3$, can be written in a condensed form called **exponent form**. For instance, the expression $3 \cdot 3 \cdot 3 \cdot 3$ can be written in exponent form as 3^4, which is read as "3 to the fourth power" or more simply as "3 to the fourth." In the expression, 3 is called the **base** and 4 is called the **exponent**. The exponent indicates the number of times the base appears as a factor. For example, 2^5 means to multiply 5 factors of 2. That is,

$$2^5 = \underbrace{2 \cdot 2 \cdot 2 \cdot 2 \cdot 2}_{5 \text{ factors of } 2}$$

EXAMPLE 1   Write in exponent form.

    (a) $4 \cdot 4$ **(b)** $(-5)(-5)(-5)(-5)(-5)(-5)(-5)$ **(c)** $7 \cdot 3 \cdot 3 \cdot 3$

SOLUTION

    (a) $4 \cdot 4 = 4^2$
    (b) $(-5)(-5)(-5)(-5)(-5)(-5)(-5) = (-5)^7$
    (c) $7 \cdot 3 \cdot 3 \cdot 3 = 7 \cdot 3^3$

To evaluate expressions containing exponents, first rewrite the expression in terms of multiplication, that is, in **expanded form**, and then calculate the product.

EXAMPLE 2   Evaluate.

    (a) 3^3 **(b)** 6^2 **(c)** $(-2)^5$ **(d)** -5^2

SOLUTION

    (a) $3^3 = 3 \cdot 3 \cdot 3 = 27$
    (b) $6^2 = 6 \cdot 6 = 36$
    (c) $(-2)^5 = (-2)(-2)(-2)(-2)(-2) = -32$
    (d) Since there are no parentheses in this expression, the base is 5 not –5. The expression can be rewritten as $-1 \cdot 5^2$

$$\begin{aligned} -5^2 &= -1 \cdot 5^2 \\ &= -1 \cdot 5 \cdot 5 \\ &= -1 \cdot 25 \\ &= -25 \end{aligned}$$

CHAPTER 1 INTEGERS   PREALGEBRA REVIEW

Powers of 10

Consider the expressions 10^2 and 10^3. Evaluating each expression gives

$$\overset{\text{Exponent is 2}}{\underset{\downarrow}{}}$$
$$10^2 = 10 \cdot 10 = 1\underset{\uparrow}{0}0$$
$$\text{Product has two zeros}$$

$$\overset{\text{Exponent is 3}}{\underset{\downarrow}{}}$$
$$10^3 = 10 \cdot 10 \cdot 10 = 1\underset{\uparrow}{0}00$$
$$\text{Product has three zeros}$$

Notice that the exponent and the number of zeros in the product are the same. This result provides a quick way to calculate powers of 10.

EXAMPLE 3   Evaluate: 10^6

SOLUTION   The exponent is 6, so the product will have a 1 followed by 6 zeros.

$10^6 = 1{,}000{,}000$

Order of Operations

Consider the expression $4 + 2 \cdot 8$.

Adding and then multiplying gives:	Multiplying and then adding gives:
$4 + 2 \cdot 8$	$4 + 2 \cdot 8$
$6 \cdot 8$	$4 + 16$
48	20

The expression appears to have two different values 48 and 20. To guarantee that an expression has a single value, an agreed upon order in which the operations are performed must be followed. This *order of operations agreement* is stated below.

PREALGEBRA REVIEW   CHAPTER 1   INTEGERS

Order of Operations
1. Do all calculations within grouping symbols, like ( ), [ ], { }, and | |, starting with the innermost grouping symbol.
2. Evaluate all expressions containing exponents.
3. Do all multiplications and divisions as they appear from left to right.
4. Do all additions and subtractions as they appear from left to right.

According to the order of operations agreement, the correct value of the expression $4 + 2 \cdot 8$ is 20.

EXAMPLE 4   Simplify: $18 - 7 + 9 - 15$

SOLUTION

$$18 - 7 + 9 - 15$$
$$11 + 9 - 15$$
$$20 - 15$$
$$5$$

EXAMPLE 5   Simplify.

(a) $4 \cdot 9 \div 12 \cdot (-3)$ (b) $21 - 6(2 + 7)$

SOLUTION

(a) $4 \cdot 9 \div 12 \cdot (-3)$
$$36 \div 12 \cdot (-3)$$
$$3 \cdot (-3)$$
$$-9$$

(b) $21 - 6(2 + 7)$
$$21 - 6(9)$$
$$21 - 54$$
$$-33$$

CHAPTER 1   INTEGERS

EXAMPLE 6   Evaluate.

(a) $24 \div 8 \cdot [10 - (5 - 1)]$ (b) $7 + |36 \div (-12 + 8)|$

SOLUTION

(a) $24 \div 8 \cdot [10 - (5 - 1)]$

$24 \div 8 \cdot [10 - 4]$

$24 \div 8 \cdot [6]$

$ 3 \cdot [6]$

$ 18$

(b) $7 + |36 \div (-12 + 8)|$

$7 + |36 \div (-4)|$

$7 + |-9|$

$7 + 9$

$ 16$

EXAMPLE 7   Simplify: $3^2 + 2^3$

SOLUTION

$$3^2 + 2^3 = 9 + 8$$
$$= 17$$

EXAMPLE 8   Evaluate: $-3(4+6)^2 + 13 - 18 \div 6$

SOLUTION

$$-3(4+6)^2 + 13 - 18 \div 6 = -3(10)^2 + 13 - 18 \div 6$$
$$= -3(100) + 13 - 18 \div 6$$
$$= -300 + 13 - 18 \div 6$$
$$= -300 + 13 - 3$$
$$= -287 - 3$$
$$= -290$$

PREALGEBRA REVIEW   CHAPTER 1 INTEGERS

In an expression such as $\dfrac{2+8\cdot 6}{1+4}$, the horizontal line between the expressions, called a fraction bar, acts as a grouping symbol. To evaluate these expressions, simplify the expression on the top and the expression on the bottom separately, then divide.

$$\frac{2+8\cdot 6}{1+4}=\frac{2+48}{5}=\frac{50}{5}=10$$

EXAMPLE 9   Simplify: $\dfrac{[(-4)^2-(-5)]+11}{9-2^4-1}$

SOLUTION

$$\begin{aligned}\frac{[(-4)^2-(-5)]+11}{9-2^4-1} &= \frac{[16-(-5)]+11}{9-16-1}\\ &= \frac{[16+5]+11}{-7-1}\\ &= \frac{[21]+11}{-8}\\ &= \frac{32}{-8}\\ &= -4\end{aligned}$$

CHAPTER 1   INTEGERS

Section 4: Exercises

Write in exponent form.

1. $6 \cdot 6 \cdot 6$
2. $9 \cdot 9 \cdot 9 \cdot 9 \cdot 9$
3. $(-7)(-7)(-7)(-7)$
4. $(-12)(-12)$
5. $8 \cdot 8 \cdot 8 \cdot 8 \cdot 3$
6. $5 \cdot (-2) \cdot (-2)$

Evaluate.

7. 2^4
8. 5^3
9. $(-1)^9$
10. $(-3)^4$
11. -4^2
12. -2^6
13. 10^4
14. 10^5

Simplify.

15. $64 \div (8)(2)$
16. $7 \times 8 + 1$
17. $10 - (-3) + 4 \cdot 5 - 9$
18. $(3)(6) - (5)(-2)$
19. $6 + 2 \cdot 7 - 13 + 1$
20. $15 + (12 - 9)$
21. $4(9 + 2)$
22. $(3 + 4) \times 5 - 13 + 7$
23. $9 - 2(12 - 8) + 16$
24. $20^2 - 10^3$
25. $5^2 \cdot (-2)^4$
26. $(4 - 1)^4$
27. $18 - 2^3 \cdot 3 \div 6 + 11$
28. $15 - |3 - 7| + 12$
29. $4 \cdot |1 - 9| + 3 \cdot 8$
30. $2 \cdot [36 \div (8 + 4)] - 6$
31. $[8 \cdot 3 \div (5 - 1)] \div 2 \cdot 3$
32. $16 \div 4 |-3 + (2 - 4)|$
33. $|(-16 + 1) + 6| \div 9 + 9$
34. $14 - 5 \cdot (2 - 4)^3 + 21$
35. $[(4 - 1)^2 + 10] - 24 \div 3 \cdot 5$
36. $\dfrac{8 \cdot 3 - 15}{12 \div 4}$
37. $\dfrac{15 - 2(8 + 7) - 19}{2 \cdot 3^3 - 5 \cdot 4}$
38. $\dfrac{45 - [8 + (12 - 5)]}{4^2 - 1}$

PREALGEBRA REVIEW   CHAPTER 1 INTEGERS

Section 5: Introduction to Averages

The **average** of a list of numbers is the sum of the numbers divided by the number of addends. For instance, the average of 4, 9, and 11 is

$$\underbrace{\frac{4+9+11}{3}}_{\text{Number of addends}}^{\text{Sum of numbers}} = \frac{24}{3} = 8$$

Note that the average is also called the *arithmetic mean*. (Mean is discussed later in the book.)

EXAMPLE 1   Some of the longest rivers in the U.S. are given below. Find the average length of the five rivers. (Source: The New York Times Almanac)

Mississippi: 3780 kilometers
Missouri: 3725 kilometers
Rio Grande: 3035 kilometers
Colorado: 2335 kilometers
Arkansas: 2335 kilometers

SOLUTION   To find the average, add the lengths of all five rivers and divide by the number of rivers in the list.

$$\frac{\text{Sum of the lengths of the rivers}}{\text{Number of rivers in list}} = \frac{3780+3725+3035+2335+2335}{5}$$
$$= \frac{15,210}{5}$$
$$= 3042$$

The average length of the five longest rivers in the U.S. is 3042 kilometers.

CHAPTER 1   INTEGERS   PREALGEBRA REVIEW

EXAMPLE 2   Wes buys four t-shirts that cost $12 each, one dress shirt that cost $32, and three polo shirts that cost $24 each. What was the average cost of each shirt?

SOLUTION   To find the average cost, find the total amount Wes spent, then divide by the number of shirts he bought. The cost and number of each type of shirt he bought is given. To find the total amount he spent, multiply and add.

Amount spent on t-shirts: $4 \cdot \$12 = \$48$
Amount spent on dress shirts: $1 \cdot \$32 = \$32$
Amount spent on polo shirts: $3 \cdot \$24 = \$72$

Total amount spent: $\$48 + \$32 + \$72 = \$152$

The average cost of each shirt is: $\dfrac{\text{Total amount spent on shirts}}{\text{Number of shirts bought}} = \dfrac{152}{8} = 19$

Each shirt cost an average of $19.

In Example 2, the steps could have been combined to find the average as follows.

$$\dfrac{\text{Total amount spent on shirts}}{\text{Number of shirts bought}} = \dfrac{4 \cdot 12 + 1 \cdot 32 + 3 \cdot 24}{8}$$
$$= \dfrac{48 + 32 + 72}{8}$$
$$= \dfrac{152}{8}$$
$$= 19$$

Finding the Unknown Value in a List

In some situations one of the values in a list of numbers is unknown, but the average of the list of numbers, including the unknown value, is known. When this occurs, the average can be used to find the unknown value in the list. To solve type of problem, first multiply the average by the number of addends. This gives the sum of all the numbers in the list. Then find the difference between this number and the sum of the known numbers. This gives the unknown number.

PREALGEBRA REVIEW CHAPTER 1 INTEGERS

EXAMPLE 3 Ada received scores of 89, 94, and 85 on the first three math exams this semester. What score must she receive on her last exam to have an average score of 90?

SOLUTION The first three exam scores and the average are given. To find the fourth exam score, first find the sum of the four exam scores by multiplying the average score by the number of addends. The number of addends is 4, since there are four exams this semester.

Average score × Number of addends   →   90 × 4

90 × 4 = 360

The sum of all the exam scores is 360. Now, subtract the sum of the first three exams from the sum of all the exam scores.

Sum of all   −   Sum of first three   →   360 − (89 + 94 + 85)
exam scores   exam scores

360 − (89 + 94 + 85) = 360 − 268 = 92

Ada must receive a 92 on the fourth exam to have an average score of 90.

EXAMPLE 4 Zack has budgeted $5 per day for lunch. This week he spends $4 on Monday, $7 on Tuesday, $5 on Wednesday, and $3 on Thursday. How much can he spend for lunch on Friday if he wants to stay within his budget?

SOLUTION The amounts he spent the first four days this week and the average amount he wants to spend each day are given.

Sum spent all the − Sum spent first   →   ($5 · 5) − ($4 + $7 + $5 + $3)
days this week   four days

(5 · 5) − (4 + 7 + 5 + 3) = 25 − 19 = 6

Zack can spend $6 for lunch on Friday.

CHAPTER 1 INTEGERS

Combined Averages

When the average of two or more groups of items is known, the average of the combined groups can be determined by combining the averages.

For instance, on a particular exam, an instructor finds that the average score was 76 for her morning class, which has 30 students, and 84 for her afternoon class, which has 18 students. Suppose she wants to know the average for both classes combined. Although it is tempting to take the average of the averages, this method is incorrect. To find the combined average, first find the sum of the scores for each class, then add those sums to get the total sum of all scores, and finally divide by the total number of students in both classes.

The sum of the scores for the morning class is the average score × the number of students: $76 \times 30 = 2280$

The sum of the scores for the afternoon class is the average score × the n umber of students: $84 \times 18 = 1512$

The total sum of all scores is the sum of the scores of morning class + Sum of scores of the afternoon class: $2280 + 1512 = 3792$

The total number of students is $30 + 18 = 48$. The average score is:

$$\frac{\text{Total sum of all scores}}{\text{Total number of students}} = \frac{3792}{48} = 79$$

The combined average score for both classes is 79.

The steps for solving this problem could have been combined as follows.

$$\begin{aligned}\frac{\text{Total sum of all scores}}{\text{Total number of students}} &= \frac{76 \times 30 + 84 \times 18}{30 + 18} \\ &= \frac{2280 + 1512}{48} \\ &= \frac{3792}{48} \\ &= 79\end{aligned}$$

PREALGEBRA REVIEW   CHAPTER 1   INTEGERS

EXAMPLE 5   On the first leg of a trip, a family drove an average of 45 miles in one hour. On the second leg of the trip, the family drove an average of 36 miles in one hour. If the family drove for 5 hours on the first leg of the trip and for 4 hours on the second leg of the trip, what was the average number of miles they drove each hour for the entire trip?

SOLUTION   To find the combined average number of miles they drove, find the total number of miles they drove, which is the sum of the number of miles driven on each leg of the trip. Then divide the sum by the total number of hours driven, $5 + 4 = 9$

Miles driven on   +   Miles driven on   → $45(5) + 36(4)$
first leg of trip   second leg of trip

$45(5) + 36(4) = 225 + 144 = 369$

The family drove a total of 369 miles. The average number of miles they drove each hour for the entire trip is:

$$\frac{\text{Total number of miles driven}}{\text{Total number of hours driven}} = \frac{369}{9} = 41$$

The family drove an average of 41 miles each hour.

CHAPTER 1   INTEGERS   PREALGEBRA REVIEW

Section 5: Exercises

Find the average of the list of numbers.

1. 2, 6

2. 8, 15, 7

3. 16, 23, 19, 28, 22, 21, 32

4. 47, 27, 24, 58

5. 3, –1, –9, 11, –14

6. –12, –8, –2, –17, 0, 3

Choose the correct answer.

7. Trisha's grocery bills last month were $56, $72, $49, and $83. What was her average grocery bill last month?

   (A) $50   (B) $260   (C) $65   (D) $52   (E) $62

8. A meteorologist keeps track of the daily high temperatures in his city. The daily high temperatures last week were 64°, 61°, 68°, 72°, 60°, 58°, and 51°. What was the average daily high temperature last week?

   (A) 60°   (B) 68°   (C) 59°   (D) 65°   (E) 62°

9. Sal studied 4 hours on Sunday, 4 hours on Monday, 5 hours on Tuesday, 6 hours on Wednesday, and 1 hour on Thursday. Find the average number of hours he studied each day.

   (A) 4   (B) 5   (C) 20   (D) 6   (E) 3

10. The quarterly profits for a small technology company are given below. Determine the company's average quarterly profit.

    Quarter 1: $418,361   Quarter 2: $296,778
    Quarter 3: $186,309   Quarter 4: –$117,848

    (A) $254,824   (B) $300,483   (C) $195,900   (D) $450,724   (E) $252,335

11. During a six week training period, an athlete rode her bike a total of 1344 miles. If the athlete trained every day, what was the average number of miles she rode her bike each day?

    (A) 32   (B) 224   (C) 45   (D) 192   (E) 38

PREALGEBRA REVIEW   CHAPTER 1 INTEGERS

12. A part-time student's income last year was $14,352. If she worked all but four weeks last year, what was her average weekly income for the weeks she worked?

   (A) $276   (B) $598   (C) $1196   (D) $299   (E) $326

13. A family bought two adult admission tickets and four child admission tickets at an amusement park. Each adult admission ticket cost $48 and each child admission ticket cost $36. What was the average cost per ticket?

   (A) $38   (B) $40   (C) $42   (D) $48   (E) $14

14. A commuter drives to work three days a week and takes the train the other two days. If it costs $25 each day he drives to work and $15 each day he takes the train, what is his average cost each day to commute to work?

   (A) $21   (B) $18   (C) $20   (D) $8   (E) $40

15. A golfer had scores of 67, 71, and 66 in his first three rounds of play at a tournament. What must he score in his last round to have an average score of 68 per round?

   (A) 65   (B) 68   (C) 67   (D) 51   (E) 69

16. A student scored 92, 87, 84, and 79 on his first four chemistry exams this semester. What score must he get on his last exam to have an average of 85 this semester?

   (A) 83   (B) 85   (C) 90   (D) 86   (E) 88

17. Gus spent $75, $51, and $47 on groceries for the first three weeks this month. How much can he spend the fourth week of the month, if he wants to stay within his budget of spending an average of $55 a week on groceries?

   (A) $73   (B) $47   (C) $52   (D) $55   (E) $50

18. Cass goes to the gym five times a week. She spent 38 minutes, 45 minutes, 60 minutes and 52 minutes at the gym the first four times she went this week. If she wants to spend an average of 45 minutes at the gym each day, how many minutes should she spend at the gym on the fifth day this week?

   (A) 39   (B) 45   (C) 48   (D) 30   (E) 56

19. If 32 is added to a group of 8 numbers whose average is 14, by how much does the average increase?

   (A) 4   (B) 3   (C) 18   (D) 16   (E) 2

CHAPTER 1   INTEGERS   PREALGEBRA REVIEW

20. A number is added to a group of 7 numbers. As a result, the average decreases from 36 to 33. What number was added?

   (A) 15   (B) 12   (C) 38   (D) –21   (E) 18

21. A student had an average of 82 on her first 3 math exams. She took a fourth exam and scored 94. What was the average of all four exams?

   (A) 85   (B) 93   (C) 89   (D) 80   (E) 86

22. The first 10 months this year, Dan spent an average of $81 a month eating out. Find the total amount he can spend the last two months of the year if he wants to keep within his budget of spending $75 a month for eating out?

   (A) $45   (B) $75   (C) $150   (D) $90   (E) $156

23. An instructor noted that there was an average of 21 students in her 3 Prealgebra classes and an average of 16 students in her 2 Intermediate Algebra classes. What was the average number of students in all her classes?

   (A) 21   (B) 9   (C) 19   (D) 17   (E) 20

24. On the first part of a trip, Jan drove for 4 hours at an average speed of 42 miles per hour (mph). On the second part of the trip, she drove 3 hours at an average speed of 56 mph. Find the average speed she drove over the entire trip.

   (A) 49 mph   (B) 50 mph   (C) 45 mph   (D) 51 mph   (E) 48 mph

25. A company had an average monthly profit of $547,854 for the first 8 months of the year. The average monthly profit for the last 4 months of the year was $268,872. What was the average monthly profit for the entire year?

   (A) $68,060   (B) $454,860   (C) $408,363   (D) $816,726   (E) $278,982

26. In a particular city, the average monthly temperature for the first 3 months of the year was 38°. The average monthly temperature for the next 6 months was 86°. What was the average monthly temperature for the last 3 months of the year if the average monthly temperature for year was 65°?

   (A) 58°   (B) 63°   (C) 62°   (D) 50°   (E) 59°

Chapter 2   Fractions

Section 1: Factors and Multiples

Factors

Recall that when multiplying two or more numbers, the numbers that are being multiplied are called *factors*. For example, in $2 \cdot 4 = 8$, both 2 and 4 are factors of 8. Note that since division is the inverse operation of multiplication, the factors of a number are those numbers by which it can be divided with no remainder. That is, since $8 \div 2 = 4$, 2 is a factor of 8, or 8 is **divisible** by 2.

> **Divisibility:** A number a is *divisible* by a number b if the number b is a factor of a. That is, when a is divided by b, the remainder is 0.

To determine whether a number is divisible by another number, divisibility tests can be used.

Tests for divisibility: A number is divisible by
- **2**, if the last digit of the number is 0, 2, 4, 6, or 8. For example, 128 is divisible by 2, since the last digit is 8. ($128 \div 2 = 64$)
- **3**, if the sum of the digits is divisible by 3. For example, 237 is divisible by 3, since the sum of the digits is $2 + 3 + 7 = 12$ and 12 is divisible by 3. ($237 \div 3 = 79$)
- **4**, if the last two digits make a number that is divisible by 4. For example, 516 is divisible by 4, since the last two digits make 16 and 16 is divisible by 4. ($516 \div 4 = 129$)
- **5**, if the last digit is 0 or 5. For example, 325 is divisible by 5, since the last digit is 5 and 320 is divisible by 5 since the last digit is 0. ($325 \div 5 = 65$ and $320 \div 5 = 64$)
- **6**, if the number is divisible by both 2 and 3. For example, 726 is divisible by 6, since it is divisible by both 2 (last digit is even) and 3 (the sum of digits is 15, which is divisible by 3). ($726 \div 6 = 121$)
- **8**, if the last three digits make a number that is divisible by 8. For example, 3232 is divisible by 8, since the last three digits make 232 and 232 is divisible by 8. ($3232 \div 8 = 404$)
- **9**, if the sum of the digits is divisible by 9. For example, 387 is divisible by 9, since the sum of the digits, $3 + 8 + 7$, or 18, is divisible by 9. ($387 \div 9 = 43$)
- **10**, if the last digit is 0. For example, 920 is divisible by 10, since the last digit is 0. ($920 \div 10 = 92$)

The expression $2 \cdot 4$ is called a *factorization* of 8. A **factorization** of a number expresses the number as a product of two or more factors. The other factorizations of the number 8 are $1 \cdot 8$ and $2 \cdot 2 \cdot 2$. Note that the last factorization has three factors. The factors of 8 are 1, 2, 4, and 8.

CHAPTER 2 FRACTIONS

EXAMPLE 1   Find all the factors of 24.

SOLUTION   To find all the factors, make a list of all the two-factor factorizations of 24.

Factorizations: $1 \cdot 24, 2 \cdot 12, 3 \cdot 8, 4 \cdot 6$
Factors: 1, 2, 3, 4, 6, 8, 12, and 24

A **prime number** is a number that has exactly two factors, 1 and the number itself. A **composite number** is a number that has more than two factors. For example, 2 is a prime number because its only factors are 1 and 2. The number 6 is a composite number because it has more than two factors, namely, 1, 2, 3 and 6. Note that the number 1 is neither a prime number nor a composite number, since its only factor is 1.

The first eight prime numbers are 2, 3, 5, 7, 11, 13, 17, and 19.

EXAMPLE 2   Determine whether the number is a prime number or a composite number.

   (a) 18   **(b)** 23   **(c)** 31   **(d)** 49

SOLUTION
   (a) composite number
   (b) prime number
   (c) prime number
   (d) composite number

A composite number can be expressed as a product of prime factors. This type of factorization is called the **prime factorization** of the number. For example, the prime factorization of 12 is $2 \cdot 2 \cdot 3$.

To find the prime factorization of a number, a factor tree and the divisibility tests can be used. For instance, the prime factorization of 24 can be found using a factor tree as follows.

$$24$$
$$\boxed{2} \cdot 12$$
$$\boxed{2} \cdot 6$$
$$\boxed{2} \cdot \boxed{3}$$

The prime factorization of 24 is $2 \cdot 2 \cdot 2 \cdot 3$.

EXAMPLE 3 Write the prime factorization of the number.

(a) 60 (b) 75

SOLUTION

(a) 60
$\wedge$
[5]·12
   $\wedge$
   [2]·6
      $\wedge$
      [2]·[3]

The prime factorization of 60 is $2 \cdot 2 \cdot 3 \cdot 5$.

(b) 75
$\wedge$
[5]·15
   $\wedge$
   [5]·[3]

The prime factorization of 75 is $3 \cdot 5 \cdot 5$.

Greatest Common Factor

The factors of 16 are 1, 2, 4, 8, and 16. The factors of 24 are 1, 2, 3, 4, 6, 8, 12, and 24. Notice that 16 and 24 have the *common factors* 1, 2, 4, and 8. The largest of these common factors, called the *greatest common factor*, is 8.

The **greatest common factor (GCF)** of two numbers is the largest number that is a factor of both numbers.

EXAMPLE 4 Find the GCF of 48 and 72.

SOLUTION Factors of 48: 1, 2, 3, 4, 6, 8, 12, 16, **24**, 48
Factors of 72: 1, 2, 3, 4, 6, 8, 9, 12, 18, **24**, 36, 72

Since 24 is the largest number that is a factor of both 48 and 72, the GCF is 24.

CHAPTER 2   FRACTIONS

The GCF can also be found from the prime factorizations. The GCF consists of all the prime factors that are common to both factorizations. Each prime factor appears in the GCF the *least* number of times it occurs in any of the factorizations. For instance, using the numbers in Example 4, the prime factorization of 48 is $2 \cdot 2 \cdot 2 \cdot 2 \cdot 3$ and the prime factorization of 72 is $2 \cdot 2 \cdot 2 \cdot 3 \cdot 3$. The prime factors common to both factorizations are 2 and 3. The least number of times 2 occurs is three and the least number of times 3 occurs is one. So the GCF of 48 and 72 is $2 \cdot 2 \cdot 2 \cdot 3$, or 24.

EXAMPLE 5   Determine the GCF of 32 and 80.

SOLUTION   Prime factorization of 32: $2 \cdot 2 \cdot 2 \cdot 2 \cdot 2$
Prime factorization of 80: $2 \cdot 2 \cdot 2 \cdot 2 \cdot 5$

The GCF is $2 \cdot 2 \cdot 2 \cdot 2 = 16$.

Multiples

A **multiple** of a number is the product of the number and a positive integer. For instance, 8 is a multiple of 2 because $8 = 2 \cdot 4$. The other multiples of 2 are 2, 4, 6, 10, 12, 14, and so on.

EXAMPLE 6   Find the first eight multiples of 4.

SOLUTION

$4 \times 1 = 4$ $4 \times 2 = 8$ $4 \times 3 = 12$ $4 \times 4 = 16$
$4 \times 5 = 20$ $4 \times 6 = 24$ $4 \times 7 = 28$ $4 \times 8 = 32$

The first eight multiples of 4 are 4, 8, 12, 16, 20, 24, 28, and 32.

EXAMPLE 7   Is 108 a multiple of 9?

SOLUTION   If 108 is a multiple of 9, then it should be divisible by 9.

$$108 \div 9 = 9\overline{)108}$$

with quotient 12 (9, 18, 18, 0).

Yes, 108 is a multiple of 9.

PREALGEBRA REVIEW   CHAPTER 2   FRACTIONS

Least Common Multiple

A *common multiple* is a number that is a multiple of two or more numbers. For instance, some common multiples of 3 and 5 are 15, 30, and 45.

EXAMPLE 8 Find three common multiples of 6 and 8.

SOLUTION Multiples of 6: 6, 12, 18, **24**, 30, 36, 42, **48**, 54, 60, 66, **72**, 78, 84, 90, ...
Multiples of 8: 8, 16, **24**, 32, 40, **48**, 56, 64, **72**, 80, ...

Three common multiples of 6 and 8 are 24, 48, and 72. Note that 6 and 8 have more than three common multiples.

The common multiples of 6 and 8 listed in Example 8 are 24, 48, and 72. The smallest of these common multiples, called the least common multiple, is 24.

The **least common multiple (LCM)** of two or more numbers is the smallest number that is a multiple of all the numbers.

EXAMPLE 9 What is the LCM of 9 and 12?

SOLUTION Multiples of 9: 9, 18, 27, **36**, 45, 54, 63, ...
Multiples of 12: 12, 24, **36**, 48, 60, ...

Since 36 is the smallest number that is a multiple of both, the LCM is 36.

EXAMPLE 10 Find the LCM of 3, 6, and 10.

SOLUTION Multiples of 3: 3, 6, 9, 12, 15, 18, 21, 24, 27, **30**, 33, ...
Multiples of 6: 6, 12, 18, 24, **30**, 36, ...
Multiples of 10: 10, 20, **30**, 40, ...

The LCM is 30.

The LCM can also be found from the prime factorizations. The LCM consists of all the different prime factors that are in each of the factorizations. Each prime factor appears in the LCM the *greatest* number of times it occurs in any of the factorizations. For instance, using the numbers in Example 9, the prime factorization of 9 is $3 \cdot 3$ and the prime factorization of 12 is $2 \cdot 2 \cdot 3$. The different prime factors in the factorizations are 2 and 3. The greatest number of times 2 occurs is two and the greatest number of times 3 occurs is two. So the LCM of 9 and 12 is $2 \cdot 2 \cdot 3 \cdot 3$, or 36.

CHAPTER 2   FRACTIONS   PREALGEBRA REVIEW

EXAMPLE 11   Find the LCM of 15 and 36.

SOLUTION   Prime factorization of 15: $3 \cdot 5$
Prime factorization of 36: $2 \cdot 2 \cdot 3 \cdot 3$

The LCM is $2 \cdot 2 \cdot 3 \cdot 3 \cdot 5 = 180$

Solving Applied Problems

EXAMPLE 12   For production purposes, the number of pages in a textbook must be a multiple of 16. Can a textbook be 496 pages long?

SOLUTION   If the textbook can be 496 pages long, then 496 is a multiple of 16. In other words, if 496 is divisible by 16, then it is a multiple of 16.

$$496 \div 16 = 16\overline{)496} \quad \begin{array}{r} 31 \\ \underline{48} \\ 16 \\ \underline{16} \\ 0 \end{array}$$

Yes, the textbook can be 496 pages long.

EXAMPLE 13   A homeowner purchases 4 inch by 6 inch decorative rectangular tiles. What are the dimensions of the smallest square she can make using these tiles?

SOLUTION   The dimensions of the rectangular tiles are given. Recall that the sides of a square have equal lengths. To find the dimensions of the smallest square she can make, find the LCM of 4 and 6.

Multiples of 4: 4, 8, **12**, 16, 20, …
Multiples of 6: 6, **12**, 18, 24, …

The LCM of 4 and 6 is 12. The smallest square she can make using these tiles has dimensions of 12 inches by 12 inches.

PREALGEBRA REVIEW   CHAPTER 2   FRACTIONS

Section 1: Exercises

List all the factors of the number.

1. 10
2. 7
3. 18
4. 32
5. 85
6. 108

Is the number a prime number or a composite number?

7. 27
8. 59
9. 83
10. 91

Write the prime factorization of the number.

11. 16
12. 39
13. 63
14. 80

Find the GCF of the numbers.

15. 16 and 64
16. 12 and 42
17. 56 and 84
18. 35 and 90
19. 96 and 144
20. 55 and 121

List the first six multiples of the number.

21. 5
22. 8

Find the LCM of the numbers.

23. 3 and 8
24. 6 and 10
25. 9 and 24
26. 28 and 56
27. 18 and 54
28. 42 and 70
29. 4, 9, and 12
30. 6, 15, and 20

CHAPTER 2   FRACTIONS   PREALGEBRA REVIEW

Choose the correct answer.

31. Which number is a multiple of 25?

    (A) 5   **(B)** 135   **(C)** 175   **(D)** 180   **(E)** 155

32. Which number is not a multiple of 16?

    (A) 80   **(B)** 144   **(C)** 124   **(D)** 112   **(E)** 160

33. If a number a is divisible by 10 and 12, then a must also be divisible by what number?

    (A) 15   **(B)** 9   **(C)** 25   **(D)** 8   **(E)** 18

34. If a number n is divisible by 6 and 27, then n is not divisible by what number?

    (A) 9   **(B)** 18   **(C)** 12   **(D)** 3   **(E)** 2

35. In the 21st century, every year that is a multiple of 4 will be a leap year. Which of the following years will be a leap year?

    (A) 2030   **(B)** 2056   **(C)** 2042   **(D)** 2066   **(E)** 2078

36. A car manufacturer recommends regularly scheduled maintenance checks every 7500 miles. Which of the following mileages would not be a regularly scheduled maintenance check?

    (A) 15,000   **(B)** 37,500   **(C)** 22,500   **(D)** 67,500   **(E)** 50,000

37. A patient has two prescriptions. One prescription is to be taken every 3 hours and the other is to be taken every 4 hours. If the patient takes both medications now, in how many hours will he again take both medications at the same time?

    (A) 6   **(B)** 24   **(C)** 8   **(D)** 12   **(E)** 18

38. A package contains 24 pieces of candy. If 16 children are attending her son's birthday party, what is the least number of packages Carla needs so that each child gets the same number of pieces in their goody bag?

    (A) 2   **(B)** 48   **(C)** 3   **(D)** 8   **(E)** 4

PREALGEBRA REVIEW

CHAPTER 2   FRACTIONS

Section 2: Introduction to Fractions

A **fraction** is a number that is the quotient of two integers. A fraction is used to represent the number of equal parts of a whole.

A fraction is of the form $\frac{a}{b}$, where b is not equal to 0. Note that the fraction has three components:

- the **numerator** (the number on top), which tells how many parts of the whole are being considered
- the **denominator** (the number on the bottom), which tells the number of equal parts of the whole is divided into
- the **fraction bar**, which separates the numerator and the denominator

For instance, in the fraction $\frac{2}{3}$, 2 is the numerator and 3 is the denominator. The fraction can be represented visually by the following diagram.

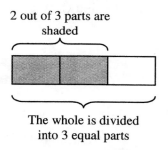

EXAMPLE 1   Use a fraction to represent the shaded portion of the figure.

(a)   (b) 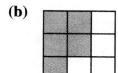   (c)

SOLUTION

(a) $\frac{1}{2}$, since 1 part out of 2 equal parts is shaded

(b) $\frac{5}{9}$, since 5 parts out of 9 equal parts are shaded

(c) $\frac{4}{4}$, since 4 parts out of 4 equal parts are shaded

Note that 1 whole figure is shaded, so this fraction is equivalent to 1.

Fractions can also be used to represent more than one whole. For example, the fraction $\frac{4}{3}$ represents more than one whole. The fraction is illustrated in the following diagram. Note that each rectangle is divided into 3 equal parts and that 4 parts are shaded.

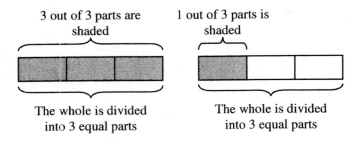

EXAMPLE 2 Write a fraction that represents the shaded portion of the figure.

SOLUTION $\frac{8}{5}$, since each rectangle is divided into 5 equal parts and 8 parts are shaded.

Proper Fractions, Improper Fractions, and Mixed Numbers

A **proper fraction** is a fraction in which the number in the numerator is smaller than the number in the denominator. The value of a proper fraction is less than 1. For instance, $\frac{3}{4}$ and $\frac{7}{12}$ are proper fractions.

An **improper fraction** is a fraction in which the number in the numerator is greater than or equal to the number in the denominator. The value of an improper fraction is greater than or equal to 1. For instance, $\frac{9}{2}$ and $\frac{5}{5}$ are improper fractions.

A **mixed number** is a number that consists of a whole number and a proper fraction. The value of a mixed number is greater than 1. For instance, $2\frac{1}{4}$, read as "two and one fourth," is a mixed number. It represents the sum of 2 wholes plus one fourth of a whole. That is, $2\frac{1}{4}$ represents the sum $2+\frac{1}{4}$.

PREALGEBRA REVIEW   CHAPTER 2   FRACTIONS

EXAMPLE 3   Identify the number as a proper fraction, an improper fraction, or a mixed number.

(a) $\dfrac{8}{8}$ (b) $7\dfrac{2}{3}$ (c) $\dfrac{11}{12}$ (d) $\dfrac{5}{4}$

SOLUTION
(a) improper fraction
(b) mixed number
(c) proper fraction
(d) improper fraction

The fraction illustrated in the following diagram can be represented in two ways: as the improper fraction $\dfrac{5}{4}$ or as the mixed number $1\dfrac{1}{4}$.

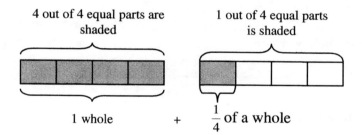

4 out of 4 equal parts are shaded   1 out of 4 equal parts is shaded

1 whole   + $\dfrac{1}{4}$ of a whole

The improper fraction $\dfrac{5}{4}$ is equivalent to the mixed number $1\dfrac{1}{4}$. This suggests that an improper fraction can be written as a mixed number or that a mixed number can be written as an improper fraction.

Recall that the fraction bar indicates division. So, $\dfrac{5}{4}$ can be written as the division problem $5 \div 4$.

> **Procedure for Writing an Improper Fraction as a Mixed Number:**
> Divide the numerator by the denominator. The quotient is the whole number part of the mixed number and the remainder is the numerator of the fraction part, which has the same denominator as the original fraction.

CHAPTER 2 FRACTIONS

EXAMPLE 4   Write $\frac{21}{8}$ as a mixed number.

SOLUTION   Divide the numerator, 21, by the denominator, 8.

$$\begin{array}{r} 2 \\ 8\overline{)21} \\ \underline{16} \\ 5 \end{array}$$

The quotient, 2, is the whole number part.

The remainder, 5, is the numerator of the fraction part, which has 8 as the denominator.

So, $\frac{21}{8}$ can be written as the mixed number $2\frac{5}{8}$.

To write a mixed number as an improper fraction the process is reversed.

Procedure for Writing a Mixed Number as an Improper Fraction:
Multiply the denominator of the fraction part times the whole number part. Then add the numerator of the fraction part to this product. The result is the numerator of the improper fraction, which has the same denominator as the fraction part in the mixed number.

EXAMPLE 5   Write $6\frac{3}{5}$ as an improper fraction.

SOLUTION   Multiply the denominator of the fraction part, 5, times the whole number part, 6. Then add the numerator of the fraction part, 3, to the product.

$$\frac{5 \times 6 + 3}{5} = \frac{30 + 3}{5} = \frac{33}{5}$$

So, $6\frac{3}{5}$ can be written as the improper fraction $\frac{33}{5}$.

PREALGEBRA REVIEW   CHAPTER 2   FRACTIONS

Negative Fractions and Mixed Numbers

Recall that the integers include both positive and negative numbers. A fraction is the quotient of two integers. Therefore, a fraction may be positive or negative. As with integers, negative fractions lie to the left of 0 on the number line and have values that are less than 0. The negative sign may be written in the numerator or denominator of a fraction. For instance, $\frac{-1}{2}$ or $\frac{1}{-2}$. However, it is generally written next to the fraction bar. For instance, $-\frac{1}{2}$. Note that, regardless of where the negative sign appears in the fraction, all three forms have the same value. That is, $\frac{-1}{2} = \frac{1}{-2} = -\frac{1}{2}$. In general, $\frac{-a}{b} = \frac{a}{-b} = -\frac{a}{b}$.

The opposite of the number in Example 5 can be written as $-6\frac{3}{5}$ or as $-\frac{33}{5}$. The negative sign does not affect the procedures for converting a mixed number and improper fractions. When converting to a negative mixed number or a negative improper fraction, ignore the negative sign for the purposes of computation, but then attach it to the answer.

EXAMPLE 6   Write $-3\frac{2}{7}$ as a single fraction.

SOLUTION

$$-3\frac{2}{7} = -\left(3\frac{2}{7}\right) = -\left(\frac{21+2}{7}\right) = -\left(\frac{23}{7}\right) = -\frac{23}{7}$$

EXAMPLE 7   Write $-\frac{17}{10}$ as a mixed number.

SOLUTION

$$10\overline{)17} \quad \rightarrow \quad \frac{17}{10} = 1\frac{7}{10}$$

So, $-\frac{17}{10} = -\left(1\frac{7}{10}\right) = -1\frac{7}{10}$

CHAPTER 2   FRACTIONS   PREALGEBRA REVIEW

Absolute Value

In Chapter 1, the absolute value of a number was defined as its distance from 0 on the number line. As with integers, the absolute value of a fraction will always be positive.

EXAMPLE 6   Find the absolute value of the number.

(a) $\left|\dfrac{3}{7}\right|$ (b) $\left|-\dfrac{5}{8}\right|$ (c) $\left|-2\dfrac{3}{4}\right|$

SOLUTION

(a) $\left|\dfrac{3}{7}\right| = \dfrac{3}{7}$

(b) $\left|-\dfrac{5}{8}\right| = \dfrac{5}{8}$

(c) $\left|-2\dfrac{3}{4}\right| = 2\dfrac{3}{4}$

PREALGEBRA REVIEW   CHAPTER 2   FRACTIONS

Section 2: Exercises

Write a fraction to represent the shaded portion of the figure.

1.

2.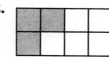

3.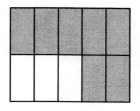

4.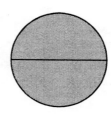

5.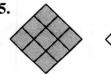

6.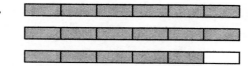

Identify the number as a proper fraction, an improper fraction, or a mixed number.

7. $\dfrac{7}{3}$

8. $\dfrac{13}{15}$

9. $\dfrac{9}{9}$

10. $1\dfrac{4}{5}$

11. $4\dfrac{2}{9}$

12. $\dfrac{24}{25}$

Write the improper fraction as an integer or mixed number.

13. $\dfrac{12}{5}$

14. $\dfrac{17}{2}$

15. $\dfrac{41}{7}$

16. $-\dfrac{53}{12}$

17. $\dfrac{18}{3}$

18. $\dfrac{76}{15}$

19. $-\dfrac{101}{10}$

20. $\dfrac{36}{9}$

21. $\dfrac{47}{-8}$

22. $\dfrac{-42}{7}$

23. $\dfrac{107}{12}$

24. $-\dfrac{133}{24}$

CHAPTER 2   FRACTIONS

Write the number as an improper fraction.

25. $1\dfrac{2}{5}$ **26.** $3\dfrac{1}{4}$ **27.** $-5\dfrac{5}{7}$ **28.** $2\dfrac{9}{11}$

29. $4\dfrac{3}{10}$ **30.** $-3\dfrac{8}{15}$ **31.** $13\dfrac{7}{9}$ **32.** $3\dfrac{19}{20}$

33. $67\dfrac{2}{3}$ **34.** $10\dfrac{3}{8}$ **35.** $-6\dfrac{4}{13}$ **36.** $-15\dfrac{5}{6}$

Determine the absolute value of the number.

37. $\left|3\dfrac{8}{9}\right|$ **38.** $\left|\dfrac{-3}{4}\right|$ **39.** $\left|\dfrac{6}{11}\right|$ **40.** $\left|-4\dfrac{1}{2}\right|$

Choose the correct answer.

41. What fraction of the figure is not shaded?

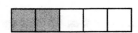

(A) $\dfrac{3}{5}$ (B) $\dfrac{5}{3}$ (C) $\dfrac{2}{5}$ (D) $\dfrac{4}{5}$ (E) $\dfrac{5}{2}$

42. What is the numerator when $7\dfrac{5}{8}$ is written as an improper fraction?

(A) 8   (B) 61   (C) 56   (D) 20   (E) 43

43. In a mathematics class of 35 students, 6 are psychology majors. What fraction of the students in the class are psychology majors?

(A) $\dfrac{35}{6}$ (B) $\dfrac{1}{6}$ (C) $\dfrac{6}{35}$ (D) $\dfrac{29}{35}$ (E) $\dfrac{3}{5}$

PREALGEBRA REVIEW

CHAPTER 2   FRACTIONS

44. Of 45 questions on an exam, 16 are multiple choice questions. What fraction of the questions on the exam are multiple choice questions?

(A) $\dfrac{29}{45}$ (B) $\dfrac{1}{45}$ (C) $\dfrac{45}{16}$ (D) $\dfrac{16}{45}$ (E) $\dfrac{1}{16}$

45. A large pizza has 16 slices. Eleven slices are eaten. What fraction of the pizza was not eaten?

(A) $\dfrac{5}{16}$ (B) $\dfrac{5}{11}$ (C) $\dfrac{11}{16}$ (D) $\dfrac{16}{11}$ (E) $\dfrac{16}{5}$

46. A typical hour long television program has 18 minutes of commercials with the rest of the time dedicated to the actual program. What fraction of the time (in minutes) is dedicated to the television program?

(A) $\dfrac{18}{60}$ (B) $\dfrac{42}{1}$ (C) $\dfrac{18}{42}$ (D) $\dfrac{1}{42}$ (E) $\dfrac{42}{60}$

47. A small company employs 27 women and 32 men. What fraction of the employees are women?

(A) $\dfrac{27}{32}$ (B) $\dfrac{32}{59}$ (C) $\dfrac{27}{59}$ (D) $\dfrac{32}{27}$ (E) $\dfrac{59}{27}$

48. A car rental company has 28 compact cars and 39 sedans. What fraction of the cars are compact cars?

(A) $\dfrac{28}{67}$ (B) $\dfrac{28}{39}$ (C) $\dfrac{39}{67}$ (D) $\dfrac{67}{28}$ (E) $\dfrac{67}{39}$

49. In a history class, 3 students are freshmen, 13 students are sophomores, and 9 students are juniors. What fraction of the students are not juniors?

(A) $\dfrac{9}{25}$ (B) $\dfrac{13}{25}$ (C) $\dfrac{3}{25}$ (D) $\dfrac{16}{25}$ (E) $\dfrac{21}{25}$

50. A landscaper plants 4 rhododendron shrubs, 8 lilac bushes, and 5 rose bushes. What fraction of the plants are not rhododendron shrubs?

(A) $\dfrac{5}{17}$ (B) $\dfrac{8}{17}$ (C) $\dfrac{13}{17}$ (D) $\dfrac{4}{17}$ (E) $\dfrac{9}{17}$

Section 3: Simplifying Fractions

Equivalent Fractions

Consider the fractions represented by the shaded portion in the following diagrams. The large rectangle in each figure is the same length. Note that, although the fractions look different, they both represent the same portion of the rectangle. In other words, the two fractions have the same value.

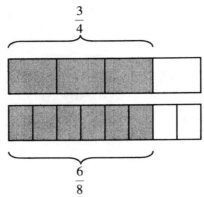

The fractions $\frac{3}{4}$ and $\frac{6}{8}$ are called *equivalent fractions*. Fractions that represent the same number are called **equivalent fractions.**

Equivalent fractions can be found by multiplying both the numerator and the denominator by the same nonzero number. For instance, some fractions that are equivalent to $\frac{3}{4}$ are:

$$\frac{3}{4} = \frac{3 \cdot 2}{4 \cdot 2} = \frac{6}{8} \qquad \frac{3}{4} = \frac{3 \cdot 3}{4 \cdot 3} = \frac{9}{12} \qquad \frac{3}{4} = \frac{3 \cdot 4}{4 \cdot 4} = \frac{12}{16}$$

EXAMPLE 1 What number must the numerator and denominator of $\frac{3}{5}$ be multiplied by to produce an equivalent fraction with a denominator of 40?

SOLUTION The denominator 5 must be multiplied by some number so that the product is 40. To find the number, divide 40 by 5.

$$40 \div 5 = 8$$

So the numerator and denominator must be multiplied by 8 to produce an equivalent fraction with a denominator of 40.

EXAMPLE 2 Write a fraction equivalent to $\dfrac{1}{2}$ with a denominator of 12.

SOLUTION Since $12 \div 2 = 6$, multiply both the numerator and denominator by 6.

$$\frac{1}{2} = \frac{1 \cdot 6}{2 \cdot 6} = \frac{6}{12}$$

EXAMPLE 3 Write a fraction equivalent to $\dfrac{7}{9}$ with a denominator of 36.

SOLUTION The numerator and denominator must be multiplied by 4.

$$\frac{7}{9} = \frac{7 \cdot 4}{9 \cdot 4} = \frac{28}{36}$$

EXAMPLE 4 Write a fraction equivalent to $\dfrac{-2}{3}$ with a denominator of 24.

SOLUTION The numerator and denominator must be multiplied by 8.

$$\frac{-2}{3} = \frac{-2 \cdot 8}{3 \cdot 8} = \frac{-16}{24} = -\frac{16}{24}$$

Note that the equivalent fraction is also negative.

EXAMPLE 5 Write a fraction equivalent to 2 with a denominator of 8.

SOLUTION Recall from Chapter 1 that $\dfrac{a}{1} = a$ for any number a. This implies that 2 can be written as a fraction with a denominator of 1: $2 = \dfrac{2}{1}$. The numerator and denominator of this fraction must be multiplied by 8 to produce an equivalent fraction with a denominator of 8.

$$\frac{2}{1} = \frac{2 \cdot 8}{1 \cdot 8} = \frac{16}{8}$$

Note that if you divide 16 by 8, the result is 2.

CHAPTER 2 FRACTIONS PREALGEBRA REVIEW

Least Common Denominator

The **least common denominator (LCD)** of two or more fractions is the least common multiple (LCM) of the denominators. For instance, the LCD of $\dfrac{2}{3}$ and $\dfrac{2}{9}$ is 9.

EXAMPLE 6 Find the LCD of each pair of fractions. Then write each fraction as an equivalent fraction with the LCD.

(a) $\dfrac{3}{4}, \dfrac{5}{6}$ (b) $\dfrac{2}{5}, \dfrac{1}{2}$ (c) $\dfrac{7}{8}, \dfrac{9}{10}$

SOLUTION

(a) The LCM of 4 and 6 is 12. So the LCD is 12. Write each fraction as an equivalent fraction with a denominator of 12.

$$\frac{3}{4} = \frac{3\cdot 3}{4\cdot 3} = \frac{9}{12} \quad \text{and} \quad \frac{5}{6} = \frac{5\cdot 2}{6\cdot 2} = \frac{10}{12}$$

(b) The LCM of 5 and 2 is 10. So the LCD is 10.

$$\frac{2}{5} = \frac{2\cdot 2}{5\cdot 2} = \frac{4}{10} \quad \text{and} \quad \frac{1}{2} = \frac{1\cdot 5}{2\cdot 5} = \frac{5}{10}$$

(c) The LCM of 8 and 10 is 40. So the LCD is 40.

$$\frac{7}{8} = \frac{7\cdot 5}{8\cdot 5} = \frac{35}{40} \quad \text{and} \quad \frac{9}{10} = \frac{9\cdot 4}{10\cdot 4} = \frac{36}{40}$$

Simplifying Fractions

A fraction is said to be in **lowest terms**, or **simplified**, if the numerator and denominator have no common factor other than 1. For instance, the fraction $\dfrac{3}{4}$ is in lowest terms because the only factor 3 and 4 have in common is 1. However, the fraction $\dfrac{9}{12}$ is not in lowest terms because 9 and 12 have a common factor of 3.

There are two ways to write a fraction in lowest terms. One way involves dividing out the GCF of the numerator and the denominator.

PREALGEBRA REVIEW   CHAPTER 2   FRACTIONS

Procedure for Writing a Fraction in Lowest Terms (using GCF):
Find the GCF of the numerator and denominator. Write the numerator and denominator as a product with the GCF as one factor. Divide out the GCF. Then multiply the remaining factors.

EXAMPLE 7   Write the fraction in lowest terms.

$$\text{(a) } \frac{12}{15} \qquad \text{(b) } \frac{48}{72} \qquad \text{(c) } -\frac{35}{42}$$

SOLUTION

(a) The GCF of 12 and 15 is 3.

$$\frac{12}{15} = \frac{3 \cdot 4}{3 \cdot 5} \qquad \textit{Write the numerator and denominator as a product with the GCF as one factor.}$$

$$= \frac{\cancel{3} \cdot 4}{\cancel{3} \cdot 5} \qquad \textit{Divide out the GCF.}$$

$$= \frac{1 \cdot 4}{1 \cdot 5} \qquad \textit{Multiply the remaining factors.}$$

$$= \frac{4}{5} \qquad \textit{Lowest terms}$$

(b) $\dfrac{48}{72} = \dfrac{24 \cdot 2}{24 \cdot 3} = \dfrac{\cancel{24} \cdot 2}{\cancel{24} \cdot 3} = \dfrac{1 \cdot 2}{1 \cdot 3} = \dfrac{2}{3}$

(c) $-\dfrac{35}{42} = -\dfrac{7 \cdot 5}{7 \cdot 6} = -\dfrac{\cancel{7} \cdot 5}{\cancel{7} \cdot 6} = -\dfrac{1 \cdot 5}{1 \cdot 6} = -\dfrac{5}{6}$

Note that the fraction written in lowest terms is also negative.

Another way to write a fraction in lowest terms involves the prime factorizations of the numerator and denominator.

Procedure for Writing a Fraction in Lowest Terms (using prime factors):
Write the prime factorization of the numerator and denominator. Divide out common prime factors. Then multiply the remaining prime factors.

CHAPTER 2 FRACTIONS

EXAMPLE 8 Simplify.

(a) $\dfrac{36}{40}$ (b) $\dfrac{21}{63}$ (c) $-\dfrac{8}{90}$

SOLUTION

(a) $\dfrac{36}{40} = \dfrac{2 \cdot 2 \cdot 3 \cdot 3}{2 \cdot 2 \cdot 2 \cdot 5}$ *Write the prime factorization of the numerator and denominator.*

$= \dfrac{\cancel{2} \cdot \cancel{2} \cdot 3 \cdot 3}{\cancel{2} \cdot \cancel{2} \cdot 2 \cdot 5}$ *Divide out the common prime factors.*

$= \dfrac{1 \cdot 1 \cdot 3 \cdot 3}{1 \cdot 1 \cdot 2 \cdot 5}$ *Multiply the remaining prime factors.*

$= \dfrac{9}{10}$ *Lowest terms*

(b) $\dfrac{21}{63} = \dfrac{3 \cdot 7}{3 \cdot 3 \cdot 7} = \dfrac{\cancel{3} \cdot \cancel{7}}{\cancel{3} \cdot 3 \cdot \cancel{7}} = \dfrac{1 \cdot 1}{1 \cdot 3 \cdot 1} = \dfrac{1}{3}$

(c) $-\dfrac{8}{90} = -\dfrac{2 \cdot 2 \cdot 2}{2 \cdot 3 \cdot 3 \cdot 5} = -\dfrac{\cancel{2} \cdot 2 \cdot 2}{\cancel{2} \cdot 3 \cdot 3 \cdot 5} = -\dfrac{1 \cdot 2 \cdot 2}{1 \cdot 3 \cdot 3 \cdot 5} = -\dfrac{4}{45}$

Comparing Fractions

Recall from Chapter 1 that a number line can be used to compare two numbers. For any two numbers on the number line, the number to the left is *less than* the number to the right. Likewise, the number to the right is the *greater than* the number to the left. The inequality symbol > means "is greater than" and the inequality symbol < means "is less than." For instance, since $\dfrac{1}{5}$ lies to the left of $\dfrac{3}{5}$ on the number line, $\dfrac{1}{5} < \dfrac{3}{5}$. Likewise, since $\dfrac{3}{5}$ lies to the right of $\dfrac{1}{5}$ on the number line, $\dfrac{3}{5} > \dfrac{1}{5}$.

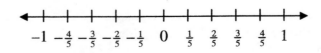

Although the number line provides a visual tool, it is not the most convenient way to compare fractions, especially when the fractions have different denominators.

PREALGEBRA REVIEW

CHAPTER 2 FRACTIONS

Procedure for Comparing Two Fractions with the Same Sign:
- *Fractions have the same denominator* – Compare the numerators. The fraction with the larger numerator is the larger fraction.
- *Fractions have different denominators* – Write the fractions so that they have the same denominator. Then compare the numerators. The fraction with the larger numerator is the larger fraction.

Recall from Chapter 1 that a positive number is always greater than a negative number. So when comparing fractions with different signs, a positive fraction is always larger than a negative fraction.

EXAMPLE 9 Write > or < between each pair of numbers to make a true statement.

(a) $\dfrac{9}{11}$ _____ $\dfrac{10}{11}$ (b) $\dfrac{2}{5}$ _____ $-\dfrac{1}{2}$

(c) $\dfrac{4}{9}$ _____ $\dfrac{5}{12}$ (d) $-\dfrac{3}{8}$ _____ $\dfrac{-1}{6}$

SOLUTION

(a) The fractions have the same denominator. Since $9 < 10$, $\dfrac{9}{11} < \dfrac{10}{11}$.

(b) $\dfrac{2}{5} > -\dfrac{1}{2}$

(c) The LCD of 9 and 12 is 36. Rewrite the fractions as equivalent fractions with the LCD and then compare the numerators.

$$\frac{4}{9} = \frac{4 \cdot 4}{9 \cdot 4} = \frac{16}{36} \quad \text{and} \quad \frac{5}{12} = \frac{5 \cdot 3}{12 \cdot 3} = \frac{15}{36}$$

Since $16 > 15$, $\dfrac{16}{36} > \dfrac{15}{36}$.

(d) The LCD of 6 and 8 is 24.

$$-\frac{3}{8} = \frac{-3}{8} = \frac{-3 \cdot 3}{8 \cdot 3} = \frac{-9}{24} \quad \text{and} \quad \frac{-1}{6} = \frac{-1 \cdot 4}{6 \cdot 4} = \frac{-4}{24}$$

Since $-9 < -4$, $\dfrac{-9}{24} < \dfrac{-4}{24}$.

CHAPTER 2 FRACTIONS   PREALGEBRA REVIEW

Section 3: Exercises

Write a fraction equivalent to the given fraction with the indicated denominator.

1. $\dfrac{1}{2}$; denominator of 18

2. $\dfrac{2}{3}$; denominator of 21

3. $\dfrac{4}{5}$; denominator of 60

4. $\dfrac{3}{4}$; denominator of 24

5. $-\dfrac{2}{7}$; denominator of 56

6. $\dfrac{-1}{6}$; denominator of 48

7. 3; denominator of 12

8. 1; denominator of 9

9. $\dfrac{7}{8}$; denominator of 32

10. $\dfrac{4}{9}$; denominator of 54

11. $\dfrac{12}{36}$; denominator of 72

12. $\dfrac{6}{18}$; denominator of 90

13. $\dfrac{-14}{25}$; denominator of 100

14. $-\dfrac{11}{16}$; denominator of 96

Find the LCD of each pair of fractions. Then write each fraction as an equivalent fraction with the LCD.

15. $\dfrac{3}{4}, \dfrac{2}{3}$

16. $\dfrac{1}{8}, \dfrac{9}{12}$

17. $\dfrac{10}{15}, \dfrac{8}{9}$

18. $\dfrac{4}{16}, \dfrac{5}{6}$

19. $\dfrac{12}{24}, \dfrac{12}{36}$

20. $\dfrac{15}{27}, \dfrac{13}{18}$

Simplify.

21. $\dfrac{8}{12}$

22. $\dfrac{6}{15}$

23. $\dfrac{10}{35}$

24. $\dfrac{18}{21}$

PREALGEBRA REVIEW   CHAPTER 2   FRACTIONS

25. $-\dfrac{16}{24}$ 26. $-\dfrac{27}{36}$ 27. $\dfrac{48}{72}$ 28. $\dfrac{32}{64}$

29. $\dfrac{25}{50}$ 30. $\dfrac{30}{80}$ 31. $\dfrac{-28}{49}$ 32. $\dfrac{-42}{56}$

33. $\dfrac{15}{22}$ 34. $\dfrac{27}{32}$ 35. $\dfrac{-77}{110}$ 36. $-\dfrac{80}{125}$

Write > or < between each pair of numbers to make a true statement.

37. $\dfrac{3}{4}$ _____ $\dfrac{1}{4}$ 38. $-\dfrac{8}{10}$ _____ $-\dfrac{3}{10}$

39. $-\dfrac{11}{24}$ _____ $\dfrac{13}{16}$ 40. $\dfrac{6}{7}$ _____ $\dfrac{-7}{12}$

41. $\dfrac{1}{2}$ _____ $\dfrac{1}{3}$ 42. $\dfrac{4}{5}$ _____ $\dfrac{13}{15}$

43. $-\dfrac{2}{8}$ _____ $-\dfrac{10}{16}$ 44. $\dfrac{11}{12}$ _____ $\dfrac{16}{18}$

Choose the correct answer.

45. What number must the numerator and denominator of $\dfrac{3}{4}$ be multiplied by to produce an equivalent fraction with a denominator of 72?

    (A) 24   (B) 16   (C) 18   (D) 13   (E) 17

46. What number must the numerator and denominator of $\dfrac{5}{8}$ be multiplied by to produce an equivalent fraction with a numerator of 80?

    (A) 16   (B) 10   (C) 12   (D) 14   (E) 18

CHAPTER 2 FRACTIONS   PREALGEBRA REVIEW

47. Which pair of fractions are equivalent?

(A) $\dfrac{12}{35}, \dfrac{18}{45}$ (B) $\dfrac{16}{22}, \dfrac{24}{33}$ (C) $\dfrac{12}{24}, \dfrac{6}{18}$ (D) $\dfrac{16}{18}, \dfrac{28}{36}$ (E) $\dfrac{8}{24}, \dfrac{20}{30}$

48. Which fraction is not equivalent to $\dfrac{9}{6}$?

(A) $\dfrac{24}{16}$ (B) $\dfrac{6}{4}$ (C) $\dfrac{21}{14}$ (D) $\dfrac{15}{20}$ (E) $\dfrac{3}{2}$

49. Which fraction is less than $\dfrac{1}{2}$?

(A) $\dfrac{3}{4}$ (B) $\dfrac{4}{7}$ (C) $\dfrac{2}{5}$ (D) $\dfrac{6}{9}$ (E) $\dfrac{8}{16}$

50. Which fraction is greater than $-\dfrac{5}{8}$?

(A) $-\dfrac{5}{6}$ (B) $\dfrac{-3}{4}$ (C) $-\dfrac{7}{9}$ (D) $-\dfrac{2}{3}$ (E) $\dfrac{-3}{5}$

51. The shaded portion of which diagram illustrates a fraction that is equivalent to $\dfrac{1}{3}$?

(A) [diagram]   (B) [diagram]   (C) [diagram]
(D) [diagram]   (E) [diagram]

52. What fraction is equivalent to the fraction illustrated by the shaded portion of the diagram?

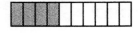

(A) $\dfrac{6}{15}$ (B) $\dfrac{3}{5}$ (C) $\dfrac{8}{25}$ (D) $\dfrac{20}{40}$ (E) $\dfrac{9}{15}$

53. What fraction of an hour is 45 minutes?

(A) $\dfrac{1}{2}$ (B) $\dfrac{3}{4}$ (C) $\dfrac{2}{3}$ (D) $\dfrac{5}{6}$ (E) $\dfrac{1}{4}$

PREALGEBRA REVIEW   CHAPTER 2   FRACTIONS

54. What fraction of an hour is 24 minutes?

   (A) $\dfrac{2}{5}$ (B) $\dfrac{1}{3}$ (C) $\dfrac{3}{8}$ (D) $\dfrac{1}{6}$ (E) $\dfrac{3}{8}$

55. What fraction of a foot is 8 inches?

   (A) $\dfrac{1}{2}$ (B) $\dfrac{3}{4}$ (C) $\dfrac{2}{3}$ (D) $\dfrac{1}{8}$ (E) $\dfrac{1}{3}$

56. What fraction of a yard is 16 inches?

   (A) $\dfrac{16}{3}$ (B) $\dfrac{4}{3}$ (C) $\dfrac{2}{9}$ (D) $\dfrac{4}{9}$ (E) $\dfrac{1}{2}$

57. The Brown family traveled 576 miles on a trip this summer. If 128 miles of their trip was by train, what fraction of their trip was by train?

   (A) $\dfrac{2}{7}$ (B) $\dfrac{7}{8}$ (C) $\dfrac{2}{3}$ (D) $\dfrac{1}{4}$ (E) $\dfrac{2}{9}$

58. Of 48 crates of apples, 18 are Red Delicious apples. What fraction of the crates are not Red Delicious apples?

   (A) $\dfrac{5}{8}$ (B) $\dfrac{3}{5}$ (C) $\dfrac{3}{8}$ (D) $\dfrac{1}{3}$ (E) $\dfrac{5}{6}$

CHAPTER 2 FRACTIONS   PREALGEBRA REVIEW

Section 4: Addition and Subtraction of Fractions

Adding and Subtracting Fractions with the Same Denominators

The following diagram illustrates the sum of the fractions $\frac{2}{7}$ and $\frac{3}{7}$.

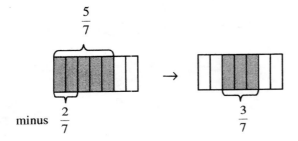

From the diagram it follows that $\frac{2}{7} + \frac{3}{7} = \frac{2+3}{7} = \frac{5}{7}$.

Subtraction of fractions can also be illustrated in a similar way. The following diagram shows the difference between $\frac{5}{7}$ and $\frac{2}{7}$.

From the diagram it follows that $\frac{5}{7} - \frac{2}{7} = \frac{5-2}{7} = \frac{3}{7}$.

Note that the fractions that were added or subtracted had the same, or common, denominators and that the denominators did not change as a result of the operations.

Procedure for Adding and Subtracting Fractions with the Same Denominators:
Add or subtract the numerators and write the result over the common denominator. Then write the sum or difference in lowest terms, if possible.

PREALGEBRA REVIEW   CHAPTER 2   FRACTIONS

EXAMPLE 1   Add or subtract. Write the answer in lowest terms.

(a) $\dfrac{1}{5}+\dfrac{3}{5}$ (b) $\dfrac{8}{9}-\dfrac{5}{9}$ (c) $\dfrac{7}{12}+\dfrac{5}{12}$ (d) $\dfrac{11}{15}-\dfrac{4}{15}$

SOLUTION

(a) $\dfrac{1}{5}+\dfrac{3}{5}=\dfrac{1+3}{5}=\dfrac{4}{5}$

(b) $\dfrac{8}{9}-\dfrac{5}{9}=\dfrac{8-5}{9}=\dfrac{3}{9}$

$=\dfrac{\cancel{3}\cdot 1}{\cancel{3}\cdot 3}=\dfrac{1\cdot 1}{1\cdot 3}=\dfrac{1}{3}$

(c) $\dfrac{7}{12}+\dfrac{5}{12}=\dfrac{7+5}{12}=\dfrac{12}{12}=1$

(d) $\dfrac{11}{15}-\dfrac{4}{15}=\dfrac{11-4}{15}=\dfrac{7}{15}$

Adding and Subtracting Fractions with Different Denominators

In order to add or subtract fractions, the fractions must have the same, or common, denominators. Recall from Section 3 that the least common denominator (LCD) of the fractions is the least common multiple (LCM) of the denominators.

> **Procedure for Adding and Subtracting Fractions with Different Denominators:**
> Use the LCD to write the fractions as equivalent fractions with a common denominator. Then add or subtract the numerators and write the result over the common denominator. Write the sum or difference in lowest terms, if possible.

CHAPTER 2   FRACTIONS

EXAMPLE 3   Add or subtract. Simplify, if possible.

(a) $\dfrac{2}{3}+\dfrac{1}{2}$ (b) $\dfrac{11}{15}-\dfrac{2}{5}$ (c) $\dfrac{8}{9}-\dfrac{3}{4}$ (d) $\dfrac{5}{6}+\dfrac{1}{8}$

SOLUTION

(a) The LCM of 3 and 2 is 6. So the LCD is 6.

$$\dfrac{2}{3}+\dfrac{1}{2}=\dfrac{2\cdot 2}{3\cdot 2}+\dfrac{1\cdot 3}{2\cdot 3}$$
$$=\dfrac{4}{6}+\dfrac{3}{6}$$
$$=\dfrac{7}{6}, \text{ or } 1\dfrac{1}{6}$$

(b) The LCM of 15 and 5 is 15. So the LCD is 15.

$$\dfrac{11}{15}-\dfrac{2}{5}=\dfrac{11}{15}-\dfrac{2\cdot 3}{5\cdot 3}$$
$$=\dfrac{11}{15}-\dfrac{6}{15}$$
$$=\dfrac{5}{15}$$
$$=\dfrac{\cancel{5}\cdot 1}{\cancel{5}\cdot 3}=\dfrac{1\cdot 1}{1\cdot 3}=\dfrac{1}{3}$$

(c) The LCM of 9 and 4 is 36. So the LCD is 36.

$$\dfrac{8}{9}-\dfrac{3}{4}=\dfrac{8\cdot 4}{9\cdot 4}-\dfrac{3\cdot 9}{4\cdot 9}=\dfrac{32}{36}-\dfrac{27}{36}=\dfrac{5}{36}$$

(d) The LCM of 6 and 8 is 24. So the LCD is 24.

$$\dfrac{5}{6}+\dfrac{1}{8}=\dfrac{5\cdot 4}{6\cdot 4}+\dfrac{1\cdot 3}{8\cdot 3}=\dfrac{20}{24}+\dfrac{3}{24}=\dfrac{23}{24}$$

PREALGEBRA REVIEW   CHAPTER 2   FRACTIONS

Adding and Subtracting Fractions, Whole Numbers, and Mixed Numbers

Recall that a whole number and a mixed number can be written as improper fractions. To write a whole number as an improper fraction, use the division fact that $\frac{a}{1} = a$ for any number a. For instance, the whole number 4 can be written as the fraction $\frac{4}{1}$.

> **Procedure for Adding and Subtracting Fractions, Whole Numbers, and Mixed Numbers:**
> Write any whole numbers and mixed numbers as improper fractions. Then add or subtract the fractions. Simplify, if possible.

EXAMPLE 4   Add.

(a) $2\frac{1}{3} + \frac{3}{4}$ (b) $4\frac{5}{6} + 2\frac{2}{9}$

SOLUTION

(a) $2\frac{1}{3} + \frac{3}{4} = \frac{7}{3} + \frac{3}{4}$ *Write the mixed number as an improper fraction.*

$= \frac{7 \cdot 4}{3 \cdot 4} + \frac{3 \cdot 3}{4 \cdot 3}$ *The LCD is 12.*

$= \frac{28}{12} + \frac{9}{12}$

$= \frac{28+9}{12} = \frac{37}{12} = 3\frac{1}{12}$

(b) $4\frac{5}{6} + 2\frac{2}{9} = \frac{29}{6} + \frac{20}{9}$

$= \frac{29 \cdot 3}{6 \cdot 3} + \frac{20 \cdot 2}{9 \cdot 2}$

$= \frac{87}{18} + \frac{40}{18}$

$= \frac{87+40}{18}$

$= \frac{127}{18} = 7\frac{1}{18}$

EXAMPLE 5 Subtract: (a) $5\dfrac{2}{5} - 4\dfrac{9}{10}$ (b) $8 - 1\dfrac{4}{11}$

SOLUTION

(a) $5\dfrac{2}{5} - 4\dfrac{9}{10} = \dfrac{27}{5} - \dfrac{49}{10}$

$= \dfrac{27 \cdot 2}{5 \cdot 2} - \dfrac{49}{10}$

$= \dfrac{54}{10} - \dfrac{49}{10} = \dfrac{54 - 49}{10} = \dfrac{5}{10}$

$= \dfrac{5 \cdot 1}{5 \cdot 2} = \dfrac{\cancel{5} \cdot 1}{\cancel{5} \cdot 2} = \dfrac{1}{2}$

(b) $8 - 1\dfrac{4}{11} = \dfrac{8}{1} - \dfrac{15}{11}$

$= \dfrac{8 \cdot 11}{1 \cdot 11} - \dfrac{15}{11} = \dfrac{88}{11} - \dfrac{15}{11} = \dfrac{88 - 15}{11} = \dfrac{73}{11} = 6\dfrac{7}{11}$

EXAMPLE 6 Add: $\dfrac{2}{3} + \dfrac{3}{5} + 1\dfrac{5}{6}$

SOLUTION

$\dfrac{2}{3} + \dfrac{3}{5} + 1\dfrac{5}{6} = \dfrac{2}{3} + \dfrac{3}{5} + \dfrac{11}{6}$

$= \dfrac{2 \cdot 10}{3 \cdot 10} + \dfrac{3 \cdot 6}{5 \cdot 6} + \dfrac{11 \cdot 5}{6 \cdot 5}$

$= \dfrac{20}{30} + \dfrac{18}{30} + \dfrac{55}{30} = \dfrac{20 + 18 + 55}{30} = \dfrac{93}{30}$

$= \dfrac{3 \cdot 31}{3 \cdot 10} = \dfrac{\cancel{3} \cdot 31}{\cancel{3} \cdot 10} = \dfrac{31}{10} = 3\dfrac{1}{10}$

Adding and Subtracting Signed Fractions and Mixed Numbers

The procedures for adding and subtracting integers work for adding and subtracting positive and negative fractions and mixed numbers.

PREALGEBRA REVIEW
CHAPTER 2 FRACTIONS

Procedure for Adding Signed Numbers:
- *Addends have the same sign* – Find the absolute values of the numbers and add them. Use the common sign as the sign of the sum. If both addends are positive, the sum is positive. If both addends are negative, the sum is negative.
- *Addends have different signs* – Find the absolute values of the numbers and subtract the smaller absolute value from the larger absolute value. Use the sign of the number with the larger absolute value as the sign of the sum.

EXAMPLE 7 Add: (a) $-\dfrac{4}{5}+\dfrac{3}{4}$ (b) $-2\dfrac{2}{3}+\left(-\dfrac{7}{9}\right)$

SOLUTION

(a) Write the fractions as equivalent fractions with the LCD of 20.

$$-\frac{4}{5}+\frac{3}{4}=-\frac{4\cdot 4}{5\cdot 4}+\frac{3\cdot 5}{4\cdot 5}=-\frac{16}{20}+\frac{15}{20}$$

Find the absolute values: $\left|-\dfrac{16}{20}\right|=\dfrac{16}{20}$ and $\left|\dfrac{15}{20}\right|=\dfrac{15}{20}$

Subtract: $\dfrac{16}{20}-\dfrac{15}{20}=\dfrac{16-15}{20}=\dfrac{1}{20}$

Since $-\dfrac{16}{20}$ has the larger absolute value and is negative, the sum is negative.

$$-\frac{4}{5}+\frac{3}{4}=-\frac{16}{20}+\frac{15}{20}=-\frac{1}{20}$$

(b) Write the mixed number as an improper fraction. Then write the fractions as equivalent fractions with the LCD of 9.

$$-2\frac{2}{3}+\left(-\frac{7}{9}\right)=-\frac{8}{3}+\left(-\frac{7}{9}\right)=-\frac{8\cdot 3}{3\cdot 3}+\left(-\frac{7}{9}\right)=-\frac{24}{9}+\left(-\frac{7}{9}\right)$$

Find the absolute values: $\left|-\dfrac{24}{9}\right|=\dfrac{24}{9}$ and $\left|-\dfrac{7}{9}\right|=\dfrac{7}{9}$

(continued on next page)

CHAPTER 2 FRACTIONS

Add: $\dfrac{24}{9} + \dfrac{7}{9} = \dfrac{31}{9} = 3\dfrac{4}{9}$

Since both addends are negative, the sum is negative.

$$-2\dfrac{2}{3} + \left(-\dfrac{7}{9}\right) = -\dfrac{8}{3} + \left(-\dfrac{7}{9}\right) = -\dfrac{24}{9} + \left(-\dfrac{7}{9}\right) = -3\dfrac{4}{9}$$

Recall that when subtracting integers, the key was to change the subtraction problem to a related addition problem. The same is true for subtracting signed fractions.

Procedure for Subtracting Signed Numbers:
To subtract $a - b$, add the opposite, or additive inverse of b to a. That is, $a - b = a + (-b)$.

EXAMPLE 8 Subtract: **(a)** $\dfrac{-8}{12} - \dfrac{2}{9}$ **(b)** $1\dfrac{1}{4} - \left(-3\dfrac{1}{6}\right)$

SOLUTION

(a) Change the subtraction problem to a related addition problem. Then proceed as usual.

$$-\dfrac{8}{12} - \dfrac{2}{9} = -\dfrac{8}{12} + \left(-\dfrac{2}{9}\right)$$

$$= -\dfrac{8}{12} + \left(-\dfrac{2}{9}\right) = -\dfrac{8 \cdot 3}{12 \cdot 3} + \left(-\dfrac{2 \cdot 4}{9 \cdot 4}\right) = -\dfrac{24}{36} + \left(-\dfrac{8}{36}\right)$$

The addends have the same sign, so the sum is negative.

$$-\dfrac{8}{12} + \left(-\dfrac{2}{9}\right) = -\dfrac{24}{36} + \left(-\dfrac{8}{36}\right) = -\dfrac{32}{36}$$

$$= -\dfrac{4 \cdot 8}{4 \cdot 9} = -\dfrac{\cancel{4} \cdot 8}{\cancel{4} \cdot 9} = -\dfrac{8}{9}$$

(b) $1\dfrac{1}{4} - \left(-3\dfrac{1}{6}\right) = 1\dfrac{1}{4} + 3\dfrac{1}{6}$

$$= \dfrac{5}{4} + \dfrac{19}{6} = \dfrac{5 \cdot 3}{4 \cdot 3} + \dfrac{19 \cdot 2}{6 \cdot 2} = \dfrac{15}{12} + \dfrac{38}{12} = \dfrac{53}{12} = 4\dfrac{5}{12}$$

PREALGEBRA REVIEW

CHAPTER 2   FRACTIONS

Solving Applied Problems

EXAMPLE 9   A contractor has $16\frac{2}{3}$ square yards of carpet to install in a room. How much is left if he uses $14\frac{3}{8}$ square yards to carpet the room?

SOLUTION   The total amount of carpet the contractor has and the amount he uses to carpet the room are given. To find the amount of carpet left, subtract.

Total amount of carpet − Amount of carpet used   → $16\frac{2}{3}$ yd² − $14\frac{3}{8}$ yd²

$$16\frac{2}{3} - 14\frac{3}{8} = \frac{50}{3} - \frac{115}{8} = \frac{50 \cdot 8}{3 \cdot 8} - \frac{115 \cdot 3}{8 \cdot 3}$$
$$= \frac{400}{24} - \frac{345}{24} = \frac{400 - 345}{24} = \frac{55}{24} = 2\frac{7}{24}$$

So $2\frac{7}{24}$ square yards of carpet are left.

EXAMPLE 10   Nate mixes $\frac{3}{16}$ gallon of vinegar to $\frac{3}{4}$ gallon of water to make a cleaning solution. How many gallons of liquid were used to make the cleaning solution?

SOLUTION   The amount of vinegar and the amount of water are given. To find the number of gallons used to make the cleaning solution, add.

Amount of vinegar + Amount of water   → $\frac{3}{16}$ gallon + $\frac{3}{4}$ gallon

$$\frac{3}{16} + \frac{3}{4} = \frac{3}{16} + \frac{3 \cdot 4}{4 \cdot 4} = \frac{3}{16} + \frac{12}{16} = \frac{3+12}{16} = \frac{15}{16}$$

Nate used $\frac{15}{16}$ gallon of liquid to make the cleaning solution.

EXAMPLE 11 As part of a remodeling project, a homeowner installs hardwood floors throughout her house. If she installed $\frac{1}{3}$ of the hardwood flooring on Thursday and another $\frac{2}{5}$ of the hardwood flooring on Friday, what fraction of the installation job remains to be completed?

SOLUTION The part of the hardwood floor installation she completed on Thursday and Friday are given. To find the fraction of the installation job that remains to be completed, first find the part of installation job she completed on Thursday and Friday combined. To do this, add.

$$\begin{array}{c}\text{Part of installation}\\ \text{job completed on}\\ \text{Thursday}\end{array} + \begin{array}{c}\text{Part of installation}\\ \text{job completed on}\\ \text{Friday}\end{array} \rightarrow \frac{1}{3}+\frac{2}{5}$$

$$\frac{1}{3}+\frac{2}{5}=\frac{1\cdot 5}{3\cdot 5}+\frac{2\cdot 3}{5\cdot 3}=\frac{5}{15}+\frac{6}{15}=\frac{5+6}{15}=\frac{11}{15}$$

Now subtract the part of the installation job already completed from 1, which represents one whole installation job.

$$\begin{array}{c}\text{Whole}\\ \text{installation job}\end{array} - \begin{array}{c}\text{Part of installation}\\ \text{job completed}\end{array} \rightarrow 1-\frac{11}{15}$$

$$1-\frac{11}{15}=\frac{1}{1}-\frac{11}{15}=\frac{1\cdot 15}{1\cdot 15}-\frac{11}{15}=\frac{15}{15}-\frac{11}{15}=\frac{15-11}{15}=\frac{4}{15}$$

So $\frac{4}{15}$ of the installation job remains to be completed by the homeowner.

EXAMPLE 12 As part of an exercise routine, Sid jogs three times a week. If she jogged $2\dfrac{1}{4}$ miles on Monday, $3\dfrac{1}{10}$ miles on Thursday, and $1\dfrac{4}{5}$ miles on Saturday, find the total number of miles she jogged over the three days.

SOLUTION The number of miles she jogged on Monday, Thursday, and Saturday are given. To find the total number of miles she jogged over the three days, add.

$$\text{Miles on Monday} + \text{Miles on Thursday} + \text{Miles on Saturday} \rightarrow 2\dfrac{1}{4} + 3\dfrac{1}{10} + 1\dfrac{4}{5}$$

$$2\dfrac{1}{4} + 3\dfrac{1}{10} + 1\dfrac{4}{5} = \dfrac{9}{4} + \dfrac{31}{10} + \dfrac{9}{5}$$

$$= \dfrac{9 \cdot 5}{4 \cdot 5} + \dfrac{31 \cdot 2}{10 \cdot 2} + \dfrac{9 \cdot 4}{5 \cdot 4}$$

$$= \dfrac{45}{20} + \dfrac{62}{20} + \dfrac{36}{20} = \dfrac{45 + 62 + 36}{20} = \dfrac{143}{20} = 7\dfrac{3}{20}$$

She jogged a total of $7\dfrac{3}{20}$ miles over the three days.

CHAPTER 2   FRACTIONS   PREALGEBRA REVIEW

Section 4: Exercises

Add. Simplify, if possible.

1. $\dfrac{3}{8}+\dfrac{7}{8}$ 2. $\dfrac{4}{11}+\dfrac{6}{11}$ 3. $\dfrac{3}{5}+\dfrac{5}{6}$ 4. $\dfrac{3}{4}+\dfrac{2}{3}$

5. $\dfrac{7}{8}+\dfrac{1}{2}$ 6. $\dfrac{1}{6}+\dfrac{13}{18}$ 7. $\dfrac{-7}{10}+\left(-\dfrac{1}{4}\right)$ 8. $\dfrac{6}{7}+\left(-\dfrac{2}{5}\right)$

9. $\dfrac{5}{8}+\dfrac{25}{36}$ 10. $\dfrac{17}{24}+\dfrac{13}{16}$ 11. $\dfrac{19}{30}+\dfrac{-13}{18}$ 12. $\dfrac{-9}{28}+\dfrac{-29}{84}$

Subtract. Simplify, if possible.

13. $\dfrac{9}{10}-\dfrac{3}{10}$ 14. $\dfrac{11}{15}-\dfrac{8}{15}$ 15. $\dfrac{3}{5}-\dfrac{1}{4}$ 16. $\dfrac{2}{3}-\dfrac{3}{7}$

17. $\dfrac{13}{16}-\dfrac{7}{8}$ 18. $\dfrac{20}{27}-\dfrac{4}{9}$ 19. $-\dfrac{11}{16}-\left(-\dfrac{5}{6}\right)$ 20. $\dfrac{19}{24}-\left(-\dfrac{3}{8}\right)$

21. $\dfrac{11}{18}-\dfrac{7}{24}$ 22. $\dfrac{25}{36}-\dfrac{9}{16}$ 23. $\dfrac{-13}{15}-\dfrac{17}{90}$ 24. $-\dfrac{17}{22}-\dfrac{6}{33}$

Add or Subtract. Simplify, if possible.

25. $4+\dfrac{1}{5}$ 26. $\dfrac{2}{3}+3$ 27. $2\dfrac{3}{4}-\dfrac{5}{6}$ 28. $1\dfrac{5}{8}-\dfrac{1}{3}$

29. $4\dfrac{8}{15}-6$ 30. $-3\dfrac{4}{9}+2$ 31. $5\dfrac{1}{12}-4\dfrac{1}{8}$ 32. $3\dfrac{3}{5}+6\dfrac{3}{10}$

33. $-8+5\dfrac{3}{4}$ 34. $-1-12\dfrac{4}{5}$ 35. $2\dfrac{5}{8}+3\dfrac{1}{6}$ 36. $5\dfrac{1}{2}+2\dfrac{2}{7}$

37. $-2\dfrac{18}{25}+1\dfrac{13}{15}$ 38. $3\dfrac{11}{20}-1\dfrac{1}{12}$ 39. $-10\dfrac{1}{5}+\left(-6\dfrac{3}{10}\right)$ 40. $-15\dfrac{5}{6}-\left(-9\dfrac{7}{12}\right)$

PREALGEBRA REVIEW CHAPTER 2 FRACTIONS

41. $-2\dfrac{4}{5}-\dfrac{8}{15}$ **42.** $\dfrac{13}{14}-\left(-1\dfrac{4}{7}\right)$ **43.** $12\dfrac{1}{2}+\left(-13\dfrac{1}{3}\right)$ **44.** $-8\dfrac{3}{5}-10\dfrac{1}{15}$

Add. Simplify, if possible.

45. $\dfrac{1}{2}+\dfrac{2}{3}+\dfrac{3}{4}$ **46.** $\dfrac{4}{5}+\dfrac{7}{8}+\dfrac{3}{10}$

47. $3\dfrac{1}{6}+\dfrac{5}{9}+2$ **48.** $1\dfrac{7}{12}+2\dfrac{1}{4}+\dfrac{3}{8}$

49. $-\dfrac{6}{7}+\dfrac{2}{3}+(-1)$ **50.** $3\dfrac{4}{9}+\left(-\dfrac{5}{6}\right)+1\dfrac{3}{4}$

Choose the correct answer.

51. If the sum of $\dfrac{13}{18}$ and $\dfrac{23}{30}$ is calculated and simplified, what is the denominator of the resulting fraction?

(A) 90   (B) 45   (C) 9   (D) 10   (E) 30

52. What is $\dfrac{9}{16}$ less than $\dfrac{15}{24}$?

(A) $-\dfrac{1}{16}$ (B) $\dfrac{5}{8}$ (C) $-\dfrac{3}{16}$ (D) $-\dfrac{1}{8}$ (E) $\dfrac{1}{16}$

53. What number must be added to $2\dfrac{3}{5}$ to produce a sum of $3\dfrac{1}{2}$?

(A) $\dfrac{9}{10}$ (B) $6\dfrac{1}{10}$ (C) $-\dfrac{9}{10}$ (D) $1\dfrac{1}{10}$ (E) $\dfrac{1}{10}$

CHAPTER 2 FRACTIONS   PREALGEBRA REVIEW

54. If $1\frac{1}{12}$ and $-\frac{7}{12}$ are both $\frac{5}{6}$ unit from a number on the number line, then what is the number?

(A) $\frac{5}{6}$ (B) $\frac{1}{2}$ (C) $\frac{3}{4}$ (D) $\frac{1}{4}$ (E) $\frac{2}{3}$

55. If two numbers are the same distance from -1 on the number line and one of the numbers is $-1\frac{5}{9}$, what is the other number?

(A) $-\frac{4}{9}$ (B) $\frac{5}{9}$ (C) $-2\frac{1}{9}$ (D) $-\frac{5}{9}$ (E) $\frac{4}{9}$

56. How much greater than $\frac{1}{2}$ is the sum of $\frac{15}{16}$ and $\frac{7}{8}$?

(A) $2\frac{5}{16}$ (B) $\frac{7}{16}$ (C) $1\frac{5}{16}$ (D) $1\frac{1}{16}$ (E) $\frac{9}{16}$

57. Alita budgets $\frac{1}{3}$ of her monthly income for rent and $\frac{1}{5}$ for utilities. What fraction of her monthly income does Alita budget for rent and utilities combined?

(A) $\frac{8}{15}$ (B) $\frac{2}{15}$ (C) $\frac{7}{15}$ (D) $\frac{1}{4}$ (E) $\frac{1}{8}$

58. How much larger than a $\frac{3}{8}$ inch wrench is a $\frac{9}{16}$ inch wrench?

(A) $\frac{2}{3}$ in.   (B) $\frac{15}{16}$ in.   (C) $\frac{1}{2}$ in.   (D) $\frac{3}{16}$ in.   (E) $\frac{3}{8}$ in.

59. A $\frac{3}{4}$ inch thick tabletop is sanded. If $\frac{1}{32}$ inch is removed during the sanding process, how thick is the tabletop after sanding?

(A) $\frac{25}{32}$ in.   (B) $\frac{1}{7}$ in.   (C) $\frac{1}{9}$ in.   (D) $\frac{23}{32}$ in.   (E) $\frac{9}{16}$ in.

60. As part of an exercise routine, Marla walks 2 miles each day. If she has already walked $\frac{4}{5}$ mile today, how many more miles does Marla need to walk?

(A) $1\frac{1}{5}$ (B) $\frac{4}{5}$ (C) $2\frac{4}{5}$ (D) $1\frac{2}{5}$ (E) $1\frac{3}{5}$

61. A plumber needs $2\frac{3}{4}$ feet of copper tubing. If he has a $4\frac{1}{2}$ feet of copper tubing, how many feet of tubing does he have left after cutting off the length he needs?

(A) $2\frac{1}{4}$ (B) $1\frac{3}{4}$ (C) $7\frac{1}{4}$ (D) $1\frac{1}{2}$ (E) $1\frac{1}{4}$

62. Ken studied $3\frac{2}{3}$ hours on Saturday and $5\frac{1}{6}$ hours on Sunday. How many more hours did he study on Sunday?

(A) $8\frac{5}{6}$ (B) $1\frac{1}{2}$ (C) $\frac{2}{3}$ (D) $3\frac{1}{3}$ (E) $1\frac{2}{3}$

63. Gayle completed $\frac{2}{9}$ of her research paper on Monday and $\frac{2}{5}$ on Tuesday. What fraction of her research paper has she completed?

(A) $\frac{2}{7}$ (B) $\frac{17}{45}$ (C) $\frac{4}{45}$ (D) $\frac{8}{45}$ (E) $\frac{28}{45}$

64. Jan and Leah jogged $6\frac{3}{10}$ miles together. If Leah jogged an additional $1\frac{5}{8}$ miles without Jan, what was the total number of miles Leah jogged?

(A) 5 (B) $1\frac{1}{4}$ (C) $7\frac{37}{40}$ (D) $7\frac{4}{9}$ (E) $6\frac{3}{10}$

CHAPTER 2 FRACTIONS   PREALGEBRA REVIEW

65. After one hour on the job, one landscaper mowed $\frac{3}{5}$ of a lawn while a second landscaper mowed $\frac{1}{4}$ of the lawn. How much of the lawn remains to be mowed by the landscapers?

(A) $\frac{4}{9}$ (B) $\frac{3}{20}$ (C) $\frac{17}{20}$ (D) $\frac{1}{2}$ (E) $\frac{1}{10}$

66. Al wants to refinish his basement. After one week, he has completed $\frac{1}{8}$ of the job. If he completes another $\frac{5}{12}$ of the job the second week, what part of the refinishing job remains?

(A) $\frac{13}{24}$ (B) $\frac{7}{24}$ (C) $\frac{1}{4}$ (D) $\frac{3}{4}$ (E) $\frac{11}{24}$

67. A spice rub recipe calls for $\frac{1}{3}$ tablespoon of cayenne pepper, 4 tablespoons of black pepper, and $1\frac{1}{2}$ tablespoons of white pepper. How many tablespoons of pepper are needed for the spice rub?

(A) $5\frac{2}{5}$ (B) $5\frac{5}{6}$ (C) 6   (D) $5\frac{3}{4}$ (E) $1\frac{5}{6}$

68. Pat buys $\frac{3}{4}$ pound of malted milk balls, $\frac{2}{3}$ pound of gum drops, and $\frac{8}{15}$ pound of caramels. How many pounds of candy did Pat buy?

(A) $\frac{39}{60}$ (B) 2   (C) $1\frac{13}{15}$ (D) $1\frac{19}{20}$ (E) $1\frac{9}{10}$

69. Jack bikes $2\frac{9}{16}$ miles from his house to his friend's house. After leaving his friend's house he bikes $\frac{5}{6}$ miles to the library. If he bikes $1\frac{3}{4}$ miles to get home, what was the total number of miles he biked today?

(A) $4\frac{9}{16}$ (B) $3\frac{17}{26}$ (C) $5\frac{7}{48}$ (D) $1\frac{7}{12}$ (E) $4\frac{43}{48}$

PREALGEBRA REVIEW

CHAPTER 2   FRACTIONS

70. A home builder purchases a $5\frac{1}{2}$ acre lot, a $7\frac{5}{8}$ acre lot, and a $1\frac{1}{3}$ acre lot. How many acres did the home builder purchase?

(A) $13\frac{7}{24}$ (B) $13\frac{7}{24}$ (C) $14\frac{13}{24}$ (D) $14\frac{11}{24}$ (E) $14\frac{7}{8}$

71. A bride decides to make her wedding invitations. Each rectangular invitation is to have a ribbon border. The invitations measure $12\frac{1}{2}$ centimeters by 15 centimeters. How much ribbon is needed for each invitation?

(A) $27\frac{1}{2}$ cm   (B) 40 cm   (C) $54\frac{1}{2}$ cm   (D) 55 cm   (E) $42\frac{1}{2}$ cm

72. A homeowner has $28\frac{1}{5}$ yards of fencing to enclose a rectangular play area for her children. If the length of the play area is $7\frac{2}{3}$ yards and the width is $5\frac{1}{4}$ yards, how many yards of fencing will she have left over after she encloses the play area?

(A) $15\frac{17}{60}$ (B) $2\frac{11}{30}$ (C) $6\frac{1}{5}$ (D) $25\frac{5}{6}$ (E) 0

73. Tara made a slipcover and pillows for her couch. She used $26\frac{1}{8}$ square yards of fabric for the slipcover and $2\frac{2}{3}$ square yards for the pillows. If she had 32 square yards of fabric, how much fabric was left after she made the slipcover and pillows?

(A) $3\frac{5}{24}$ (B) $3\frac{8}{11}$ (C) $4\frac{5}{24}$ (C) $28\frac{19}{24}$ (E) $3\frac{5}{6}$

74. After graduation, Hank drove from school to his home, stopping twice along the way. He drove $2\frac{3}{4}$ hours on the first part of the trip and $3\frac{1}{6}$ hours on the second part of the trip. If he drove a total of 9 hours, how many hours did he drive on the third part of the trip?

(A) $5\frac{11}{12}$ (B) $3\frac{5}{6}$ (C) $3\frac{3}{5}$ (D) $4\frac{1}{12}$ (E) $3\frac{1}{12}$

CHAPTER 2 FRACTIONS PREALGEBRA REVIEW

Section 5: Multiplication and Division of Fractions

Multiplication of Fractions

Consider a rectangle that is divided into 4 equal pieces, 3 of which are shaded. That is, $\frac{3}{4}$ of the rectangle is shaded.

If $\frac{1}{2}$ of the $\frac{3}{4}$ is shaded, the original shaded portion is divided into 2 equal parts, 1 of which is shaded; and the original rectangle is divided into 8 equal pieces, 3 of which are shaded twice.

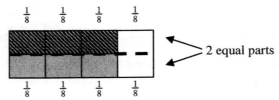

The above diagrams illustrates multiplication of the fraction $\frac{3}{4}$ by $\frac{1}{2}$. So, $\frac{1}{2}$ of $\frac{3}{4}$ is $\frac{3}{8}$. That is, $\frac{1}{2} \times \frac{3}{4} = \frac{3}{8}$. Notice that the numerator of the product, 3, is the product of the numerators of the factors and that the denominator of the product, 8, is the product of the denominators of the factors.

> **Procedure for Multiplying Fractions:**
> Multiply the numerators and multiply the denominators. Simplify the product, if possible.

EXAMPLE 1 Multiply: $\frac{2}{3} \times \frac{5}{7}$

SOLUTION

$$\frac{2}{3} \times \frac{5}{7} = \frac{2 \times 5}{3 \times 7} = \frac{10}{21}$$

Note that the product is in lowest terms.

PREALGEBRA REVIEW

CHAPTER 2 FRACTIONS

In the preceding example, the product was in lowest terms. However, often the product of two fractions is not in lowest terms. Writing the prime factorizations of the factors in the numerator and the factors in the denominator and dividing out the common factors before multiplying usually saves time. If all the common factors are divided out, the product will be in lowest terms.

EXAMPLE 2 Multiply.

(a) $\dfrac{4}{5} \cdot \dfrac{15}{16}$ (b) $\dfrac{7}{8} \times \dfrac{2}{9}$ (c) $\dfrac{12}{25}\left(\dfrac{10}{27}\right)$

SOLUTION

(a) $\dfrac{4}{5} \cdot \dfrac{15}{16} = \dfrac{4 \cdot 15}{5 \cdot 16} = \dfrac{2 \cdot 2 \cdot 3 \cdot 5}{5 \cdot 2 \cdot 2 \cdot 2 \cdot 2} = \dfrac{\cancel{2} \cdot \cancel{2} \cdot 3 \cdot \cancel{5}}{\cancel{5} \cdot \cancel{2} \cdot \cancel{2} \cdot 2 \cdot 2} = \dfrac{3}{4}$

(b) $\dfrac{7}{8} \times \dfrac{2}{9} = \dfrac{7 \times 2}{8 \times 9} = \dfrac{7 \times 2}{2 \times 2 \times 2 \times 3 \times 3} = \dfrac{7 \times \cancel{2}}{\cancel{2} \times 2 \times 2 \times 3 \times 3} = \dfrac{7}{36}$

(c) $\dfrac{12}{25}\left(\dfrac{10}{27}\right) = \dfrac{12 \cdot 10}{25 \cdot 27} = \dfrac{2 \cdot 2 \cdot 3 \cdot 2 \cdot 5}{5 \cdot 5 \cdot 3 \cdot 3 \cdot 3} = \dfrac{2 \cdot 2 \cdot \cancel{3} \cdot 2 \cdot \cancel{5}}{\cancel{5} \cdot 5 \cdot \cancel{3} \cdot 3 \cdot 3} = \dfrac{8}{45}$

EXAMPLE 3 Multiply: $\dfrac{1}{10} \cdot \dfrac{5}{9} \cdot \dfrac{2}{3}$

SOLUTION

$\dfrac{1}{10} \cdot \dfrac{5}{9} \cdot \dfrac{2}{3} = \dfrac{1 \cdot 5 \cdot 2}{10 \cdot 9 \cdot 3} = \dfrac{1 \cdot 5 \cdot 2}{2 \cdot 5 \cdot 3 \cdot 3 \cdot 3} = \dfrac{1 \cdot \cancel{5} \cdot \cancel{2}}{\cancel{2} \cdot \cancel{5} \cdot 3 \cdot 3 \cdot 3} = \dfrac{1}{27}$

Reciprocals

Two numbers are **reciprocals** of each other if their product is 1. For instance, $\dfrac{2}{3}$ and $\dfrac{3}{2}$ are reciprocals because $\dfrac{2}{3} \cdot \dfrac{3}{2} = \dfrac{6}{6} = 1$.

CHAPTER 2   FRACTIONS

To find the reciprocal of a fraction, interchange the numerator and the denominator.

EXAMPLE 3   Find the reciprocal.

(a) $\dfrac{2}{5}$ (b) $\dfrac{1}{6}$ (c) $\dfrac{7}{4}$

SOLUTION

(a) $\dfrac{5}{2}$ (b) $\dfrac{6}{1} = 6$ (c) $\dfrac{4}{7}$

Note that 0 does not have a reciprocal because there is no number that 0 can be multiplied by that results in a product of 1.

Division of Fractions

To divide two fractions, rewrite the division problem as a related multiplication problem using the reciprocal of the divisor. For example, $\dfrac{3}{8} \div \dfrac{3}{4}$ can be rewritten as $\dfrac{3}{8} \cdot \dfrac{4}{3}$.

Procedure for Dividing Two Fractions:
Multiply the dividend by the reciprocal of the divisor. Simplify the quotient, if possible.

EXAMPLE 4   Divide.

(a) $\dfrac{9}{10} \div \dfrac{3}{5}$ (b) $\dfrac{1}{12} \div \dfrac{6}{7}$

SOLUTION

(a) $\dfrac{9}{10} \div \dfrac{3}{5} = \dfrac{9}{10} \cdot \dfrac{5}{3} = \dfrac{9 \cdot 5}{10 \cdot 3} = \dfrac{3 \cdot 3 \cdot 5}{2 \cdot 5 \cdot 3} = \dfrac{\cancel{3} \cdot 3 \cdot \cancel{5}}{2 \cdot \cancel{5} \cdot \cancel{3}} = \dfrac{3}{2}$, or $1\dfrac{1}{2}$

(b) $\dfrac{1}{12} \div \dfrac{6}{7} = \dfrac{1}{12} \cdot \dfrac{7}{6} = \dfrac{1 \cdot 7}{12 \cdot 6} = \dfrac{7}{72}$

PREALGEBRA REVIEW   CHAPTER 2   FRACTIONS

Multiplying and Dividing Fractions, Whole Numbers, and Mixed Numbers

As with addition and subtraction, write any whole numbers and mixed numbers as improper fractions before multiplying and dividing.

> **Procedure for Multiplying and Dividing Fractions, Whole Numbers, and Mixed Numbers:**
> Write any whole numbers and mixed numbers as improper fractions. Then multiply or divide the fractions. Simplify, if possible.

EXAMPLE 5   Multiply: **(a)** $2\dfrac{1}{2} \times \dfrac{8}{15}$ **(b)** $1\dfrac{2}{3} \cdot 4\dfrac{1}{5}$

SOLUTION

(a) $2\dfrac{1}{2} \times \dfrac{8}{15} = \dfrac{5}{2} \times \dfrac{8}{15}$

$= \dfrac{5 \times 8}{2 \times 15} = \dfrac{5 \times 2 \times 2 \times 2}{2 \times 3 \times 5} = \dfrac{\cancel{5} \times \cancel{2} \times 2 \times 2}{\cancel{2} \times 3 \times \cancel{5}} = \dfrac{4}{3}, \text{ or } 1\dfrac{1}{3}$

(b) $1\dfrac{2}{3} \cdot 4\dfrac{1}{5} = \dfrac{5}{3} \cdot \dfrac{21}{5}$

$= \dfrac{5 \cdot 21}{3 \cdot 5} = \dfrac{5 \cdot 3 \cdot 7}{3 \cdot 5} = \dfrac{\cancel{5} \cdot \cancel{3} \cdot 7}{\cancel{3} \cdot \cancel{5}} = \dfrac{7}{1} = 7$

EXAMPLE 6   Divide: **(a)** $\dfrac{5}{6} \div 4$ **(b)** $3\dfrac{3}{8} \div 5\dfrac{1}{7}$

SOLUTION

(a) $\dfrac{5}{6} \div 4 = \dfrac{5}{6} \div \dfrac{4}{1}$

$= \dfrac{5}{6} \cdot \dfrac{1}{4} = \dfrac{5}{24}$

(b) $3\dfrac{3}{8} \div 5\dfrac{1}{7} = \dfrac{27}{8} \div \dfrac{36}{7}$

$= \dfrac{27}{8} \cdot \dfrac{7}{36} = \dfrac{3 \cdot 3 \cdot 3 \cdot 7}{2 \cdot 2 \cdot 2 \cdot 2 \cdot 2 \cdot 3 \cdot 3} = \dfrac{\cancel{3} \cdot \cancel{3} \cdot 3 \cdot 7}{2 \cdot 2 \cdot 2 \cdot 2 \cdot 2 \cdot \cancel{3} \cdot \cancel{3}} = \dfrac{21}{32}$

CHAPTER 2 FRACTIONS   PREALGEBRA REVIEW

Multiplying and Dividing Signed Fractions

The sign rules for multiplying and dividing integers work for multiplying and dividing fractions.

> **Sign Rules for Multiplying (or Dividing) Two Signed Numbers:**
> - *Same sign* – If the two numbers have the same sign, then the product (or quotient) is positive.
> - *Different signs* – If the two numbers have different signs, then the product (or quotient) is negative.

EXAMPLE 7 Multiply.

$$\text{(a)} \left(-\frac{7}{10}\right)\left(\frac{16}{21}\right) \qquad \text{(b)} \ -2\frac{5}{6}\cdot(-2)$$

SOLUTION

(a) Since the factors have different signs, the product is negative.

$$\left(-\frac{7}{10}\right)\left(\frac{16}{21}\right) = -\frac{7\cdot 16}{10\cdot 21} = -\frac{\cancel{7}\cdot\cancel{2}\cdot 2\cdot 2\cdot 2}{\cancel{2}\cdot 5\cdot 3\cdot \cancel{7}} = -\frac{8}{15}$$

(b) Since both factors have the same sign, the product is positive.

$$-2\frac{5}{6}\cdot(-2) = -\frac{17}{6}\cdot\left(-\frac{2}{1}\right) = \frac{17\cdot 2}{6\cdot 1} = \frac{17\cdot\cancel{2}}{\cancel{2}\cdot 3\cdot 1} = \frac{17}{3}, \text{ or } 5\frac{2}{3}$$

EXAMPLE 8 Divide.

$$\text{(a)} \ -\frac{3}{4}\div\left(-\frac{33}{40}\right) \qquad \text{(b)} \ -1\frac{5}{9}\div 4\frac{1}{3}$$

SOLUTION

(a) $-\dfrac{3}{4}\div\left(-\dfrac{33}{40}\right) = -\dfrac{3}{4}\cdot\left(-\dfrac{40}{33}\right) = \dfrac{3\cdot 40}{4\cdot 33} = \dfrac{\cancel{3}\cdot\cancel{2}\cdot\cancel{2}\cdot 2\cdot 5}{\cancel{2}\cdot\cancel{2}\cdot\cancel{3}\cdot 11} = \dfrac{10}{11}$

(b) $-1\dfrac{5}{9}\div 4\dfrac{1}{3} = -\dfrac{14}{9}\div\dfrac{13}{3} = -\dfrac{14}{9}\cdot\dfrac{3}{13} = -\dfrac{14\cdot 3}{9\cdot 13} = -\dfrac{2\cdot 7\cdot\cancel{3}}{\cancel{3}\cdot 3\cdot 13} = -\dfrac{14}{39}$

Solving Applied Problems

EXAMPLE 9 Kareem puts $\frac{1}{5}$ of his monthly pay into his savings account. If his monthly pay this month was $1785, how much money did he put into his savings account?

SOLUTION Kareem's monthly pay and the fraction of his pay he puts into savings are given. To find the amount he put into savings, multiply.

Fraction of pay put into savings × Monthly pay → $\frac{1}{5} \times \$1785$

$$\frac{1}{5} \times 1785 = \frac{1}{5} \times \frac{1785}{1} = \frac{1 \times 1785}{5 \times 1} = \frac{3 \times \cancel{5}^{1} \times 119}{\cancel{5}_{1} \times 1} = \frac{357}{1} = 357$$

Kareem put $357 into his savings account this month.

EXAMPLE 10 A rectangular garden measures $4\frac{1}{6}$ yards by $3\frac{3}{5}$ yards. What is the area of the garden?

SOLUTION The dimensions of the garden are given. To find the area multiply the length and width.

Length × Width → $4\frac{1}{6}$ yards × $3\frac{3}{5}$ yards

$$4\frac{1}{6} \times 3\frac{3}{5} = \frac{25}{6} \times \frac{18}{5} = \frac{25 \times 18}{6 \times 5} = \frac{\cancel{5} \times 5 \times \cancel{2} \times \cancel{3} \times 3}{\cancel{2} \times \cancel{3} \times \cancel{5}} = \frac{15}{1} = 15$$

The area of the garden is 15 square yards.

CHAPTER 2 FRACTIONS   PREALGEBRA REVIEW

EXAMPLE 11 According to the nutrition label on a box of cereal, one serving is $\frac{3}{4}$ cups. How many servings are in a box that has $17\frac{1}{2}$ cups of cereal?

SOLUTION The amount in one serving and the total amount of cereal are given. To find the number of servings in a box of cereal, divide.

Total amount of cereal ÷ Amount in one serving → $17\frac{1}{2}$ cups ÷ $\frac{3}{4}$ cups

$$17\frac{1}{2} \div \frac{3}{4} = \frac{35}{2} \div \frac{3}{4} = \frac{35}{2} \times \frac{4}{3} = \frac{35 \times 4}{2 \times 3} = \frac{5 \times 7 \times \cancel{2}^{1} \times 2}{\cancel{2}_{1} \times 3} = \frac{70}{3} = 23\frac{1}{3}$$

There are $23\frac{1}{3}$ servings in the box of cereal.

EXAMPLE 12 A carpenter is making shelves for a book case. If each shelf is $1\frac{2}{3}$ feet long, how many shelves can she cut from a board that is 8 feet long?

SOLUTION Both the length of the shelf and the length of the board are given. To find how many shelves she can cut, divide.

Length of board ÷ Length of shelf → 8 feet ÷ $1\frac{2}{3}$ feet

$$8 \div 1\frac{2}{3} = \frac{8}{1} \div \frac{5}{3} = \frac{8}{1} \cdot \frac{3}{5} = \frac{8 \cdot 3}{1 \cdot 5} = \frac{24}{5} = 4\frac{4}{5}$$

She can cut 4 shelves from the board. Note that the fractional part of the mixed number represents a fractional portion of a shelf and is not considered as part of the answer.

In Example 12, the fractional portion in the answer also represents the amount of the board that is left over. That is, $\frac{4}{5} \cdot 1\frac{2}{3} = \frac{4}{5} \cdot \frac{5}{3} = \frac{4 \cdot \cancel{5}^{1}}{\cancel{5}_{1} \cdot 3} = \frac{4}{3}$, or $1\frac{1}{3}$ feet of board is left over.

PREALGEBRA REVIEW

CHAPTER 2   FRACTIONS

Section 5: Exercises

Multiply. Simplify, if possible.

1. $\dfrac{1}{2} \cdot \dfrac{1}{10}$
2. $\dfrac{4}{5} \cdot \dfrac{1}{3}$
3. $\dfrac{6}{11}\left(\dfrac{5}{12}\right)$
4. $\left(\dfrac{15}{16}\right)\left(\dfrac{8}{9}\right)$

5. $\dfrac{5}{6} \cdot \dfrac{6}{25}$
6. $\dfrac{27}{28} \times \dfrac{4}{9}$
7. $-\dfrac{2}{3} \cdot \dfrac{1}{8}$
8. $\left(-\dfrac{4}{9}\right)\left(-\dfrac{12}{13}\right)$

9. $4 \cdot \dfrac{9}{16}$
10. $\dfrac{7}{12} \cdot 8$
11. $-1\dfrac{5}{6}\left(-\dfrac{20}{33}\right)$
12. $5\dfrac{7}{9}\left(-\dfrac{9}{14}\right)$

13. $3\dfrac{3}{4} \cdot \dfrac{2}{5}$
14. $\dfrac{28}{45}\left(1\dfrac{2}{7}\right)$
15. $-65 \cdot 2\dfrac{6}{13}$
16. $-1\dfrac{7}{16} \cdot 48$

17. $4\dfrac{2}{5} \times 7\dfrac{1}{2}$
18. $7\dfrac{1}{5} \cdot 2\dfrac{2}{9}$
19. $-8\dfrac{3}{4} \cdot 5\dfrac{5}{7}$
20. $\left(-3\dfrac{5}{8}\right)\left(-1\dfrac{1}{7}\right)$

Find the reciprocal.

21. $\dfrac{1}{3}$
22. $\dfrac{8}{11}$
23. $\dfrac{16}{9}$
24. $\dfrac{15}{4}$

25. 4
26. 2
27. $-\dfrac{10}{11}$
28. $-\dfrac{21}{12}$

Divide. Simplify, if possible.

29. $\dfrac{1}{5} \div \dfrac{1}{4}$
30. $\dfrac{3}{5} \div \dfrac{2}{3}$
31. $\dfrac{7}{9} \div \dfrac{5}{6}$
32. $\dfrac{8}{15} \div \dfrac{8}{21}$

33. $-\dfrac{12}{25} \div \dfrac{3}{20}$
34. $\dfrac{7}{18} \div \left(-\dfrac{14}{27}\right)$
35. $\dfrac{-4}{5} \div \left(-\dfrac{16}{35}\right)$
36. $\dfrac{15}{28} \div \dfrac{45}{56}$

37. $2 \div \dfrac{3}{8}$
38. $\dfrac{5}{6} \div 10$
39. $\dfrac{4}{9} \div 1\dfrac{1}{3}$
40. $1\dfrac{2}{9} \div \dfrac{11}{18}$

CHAPTER 2 FRACTIONS   PREALGEBRA REVIEW

41. $-3\dfrac{3}{4} \div \dfrac{15}{16}$ 42. $-\dfrac{3}{10} \div \left(-2\dfrac{2}{5}\right)$ 43. $4\dfrac{1}{2} \div 6$ 44. $9 \div 5\dfrac{1}{4}$

45. $-1\dfrac{3}{5} \div \left(-2\dfrac{6}{7}\right)$ 46. $-4\dfrac{5}{8} \div 2\dfrac{13}{16}$ 47. $5\dfrac{4}{9} \div 2\dfrac{1}{3}$ 48. $3\dfrac{7}{15} \div 1\dfrac{1}{12}$

Choose the correct answer.

49. What is the product of $\dfrac{2}{3}$ and $\dfrac{6}{9}$ in simplest form?

   (A) $\dfrac{2}{3}$ (B) 1   (C) $\dfrac{8}{27}$ (D) $\dfrac{4}{9}$ (E) $1\dfrac{1}{3}$

50. Divide $\dfrac{7}{10}$ by $\dfrac{28}{35}$. The quotient in lowest terms is

   (A) $\dfrac{14}{25}$ (B) $\dfrac{7}{8}$ (C) 2   (D) $1\dfrac{1}{7}$ (E) $\dfrac{1}{14}$

51. The product of a number and $\dfrac{3}{5}$ is $\dfrac{1}{2}$. What is the number?

   (A) $\dfrac{3}{10}$ (B) $1\dfrac{1}{5}$ (C) $\dfrac{5}{6}$ (D) $\dfrac{1}{3}$ (E) $1\dfrac{2}{5}$

52. If $4\dfrac{2}{3}$ is divided by a number, the quotient is $\dfrac{2}{3}$. What is the number?

   (A) 7   (B) 4   (C) $2\dfrac{1}{3}$ (D) $3\dfrac{1}{9}$ (E) $\dfrac{1}{7}$

53. The reciprocal of a number times $\dfrac{5}{6}$ is 5. What is the number?

   (A) 6   (B) $\dfrac{6}{5}$ (C) $\dfrac{1}{5}$ (D) $\dfrac{5}{6}$ (E) $\dfrac{1}{6}$

PREALGEBRA REVIEW

CHAPTER 2 FRACTIONS

54. The product of $-\dfrac{8}{9}$ and its reciprocal is

(A) –1 (B) $\dfrac{64}{81}$ (C) 1 (D) $-\dfrac{64}{81}$ (E) 0

55. At the first showing of a new movie, the theater was $\dfrac{4}{5}$ full. If the theater has 280 seats, how many people attended the first showing?

(A) 224 (B) 250 (C) 350 (D) 212 (E) 244

56. Of 348 voluntary surveys sent out to employees, only $\dfrac{1}{6}$ of the employees responded. How many employees responded?

(A) 290 (B) 116 (C) 58 (D) 72 (E) 68

57. A supermarket pre-packages Swiss cheese. How many $\dfrac{3}{4}$ pound packages can be made from a 12 pound block of Swiss cheese?

(A) 12 (B) 6 (C) 9 (D) 18 (E) 16

58. A $\dfrac{5}{8}$ meter piece of wire is to be cut into 10 pieces of equal length. How many meters long is each piece?

(A) $6\dfrac{1}{4}$ (B) 16 (C) $\dfrac{1}{4}$ (D) $\dfrac{1}{16}$ (E) $\dfrac{1}{12}$

59. A daycare center fenced off a $4\dfrac{1}{2}$ yard by $3\dfrac{2}{3}$ yard rectangular area for a play yard. What is the area of the play yard?

(A) $12\dfrac{1}{3}$ yd² (B) $16\dfrac{1}{2}$ yd² (C) $16\dfrac{1}{3}$ yd² (D) $8\dfrac{1}{6}$ yd² (E) $1\dfrac{5}{22}$ yd²

CHAPTER 2 FRACTIONS   PREALGEBRA REVIEW

60. Jack wants to build an 800 square yard rectangular dog run behind his house. What is the width of the dog run if the length is $33\frac{1}{3}$ yards?

(A) 24 yd   (B) 18 yd   (C) $24\frac{8}{33}$ yd   (D) 30 yd   (E) $22\frac{2}{3}$ yd

61. A marathon runner ran $3\frac{1}{2}$ kilometers in $15\frac{3}{4}$ minutes. How many kilometers did she run in one minute?

(A) $4\frac{1}{2}$ (B) $55\frac{1}{8}$ (C) $\frac{2}{9}$ (D) $1\frac{1}{3}$ (E) $\frac{3}{4}$

62. A recipe calls for $1\frac{2}{3}$ cups of sugar. How much sugar is needed if the recipe is doubled?

(A) $\frac{5}{6}$ (B) $3\frac{1}{3}$ (C) $2\frac{2}{3}$ (D) 3   (E) $3\frac{2}{3}$

63. A patient is prescribed $1\frac{1}{2}$ tablets of medicine to be taken three times per day for two weeks. What is the total number of tablets he takes in two weeks?

(A) 9   (B) 18   (C) 42   (D) 63   (E) 45

64. Elena contributes $\frac{2}{25}$ of her monthly salary to a 401K plan. If her monthly salary is $2275, how much does she contribute to her 401K plan in one year?

(A) $182   (B) $2500   (C) $2368   (D) $2275   (E) $2184

65. One serving of chocolate ice cream contains 16 grams of sugar. If one serving is $\frac{1}{2}$ cup, then how much sugar is in $2\frac{1}{4}$ cups of chocolate ice cream?

(A) 18 g   (B) 72 g   (C) 36 g   (D) 48 g   (E) 64 g

66. Of the 60 students in a chemistry class, $\frac{2}{5}$ are women and $\frac{5}{8}$ of these students are in the nursing program. How many of the women in the chemistry class are not in the nursing program?

(A) 15 (B) 6 (C) 9 (D) 12 (E) 24

67. Frank worked $39\frac{3}{5}$ hours over the past four days. If he worked the same number of hours each day, how many hours did he work each day?

(A) $9\frac{1}{9}$ (B) $8\frac{3}{4}$ (C) $8\frac{4}{5}$ (D) $9\frac{9}{10}$ (E) $9\frac{1}{5}$

68. A bottle is $\frac{1}{3}$ empty. If the bottle contains $2\frac{2}{3}$ quarts of liquid, what is the total capacity of the bottle?

(A) $3\frac{1}{3}$ q (B) 4 q (C) 8 q (D) 3 q (E) 6 q

69. For a craft project, Lena cut a piece of ribbon into 6 pieces of equal length. If each piece was $3\frac{1}{2}$ inches long and 3 inches of ribbon were left over, how long was the original piece of ribbon?

(A) 21 in (B) 39 in (C) 36 in (D) 24 in (E) 63 in

70. Each window valence requires $1\frac{5}{8}$ yards of fabric. If Jill has $15\frac{1}{2}$ yards of fabric, how much fabric does she have left after making 9 valences?

(A) $1\frac{5}{8}$ yd (B) $1\frac{13}{18}$ yd (C) $\frac{7}{8}$ yd (D) $\frac{1}{8}$ yd (E) $1\frac{3}{8}$ yd

71. A shirt requires $1\frac{1}{9}$ yards of fabric. If each yard of fabric costs $8 per yard, how much will it cost to make 45 shirts?

(A) $400 (B) $324 (C) $380 (D) $360 (E) $384

CHAPTER 2 FRACTIONS — PREALGEBRA REVIEW

72. A home builder bought a $15\frac{2}{3}$ acre parcel of land. If she sets aside $2\frac{1}{3}$ acres for a playground and uses the rest for $\frac{5}{6}$ acre home lots, how many home lots will there be in the development?

    (A) 18   **(B)** 16   **(C)** $10\frac{5}{6}$ **(D)** 8   **(E)** 14

73. Marina has had three children. Their birth weights were $8\frac{5}{16}$ pounds, $7\frac{1}{2}$ pounds, and $7\frac{5}{8}$ pounds. What was the average weight of her babies?

    (A) $7\frac{3}{4}$ lb   **(B)** $7\frac{9}{16}$ lb   **(C)** $7\frac{15}{16}$ lb   **(D)** $7\frac{13}{16}$ lb   **(E)** $7\frac{7}{8}$ lb

74. An athlete in training ran $8\frac{1}{2}$ miles on Monday, $10\frac{2}{5}$ miles on Wednesday, 15 miles on Friday, and $9\frac{5}{6}$ miles on Saturday. What was the average number if miles she ran each day?

    (A) $12\frac{8}{15}$ **(B)** $10\frac{14}{15}$ **(C)** $11\frac{19}{30}$ **(D)** $14\frac{26}{45}$ **(E)** $10\frac{3}{5}$

75. Evie studied an average of $4\frac{1}{6}$ hours over the past three days. If she studied $5\frac{3}{4}$ hours today, what was the average number of hours she studied for the four days?

    (A) $4\frac{5}{6}$ **(B)** $4\frac{9}{16}$ **(C)** $4\frac{1}{4}$ **(D)** $4\frac{23}{24}$ **(E)** 5

76. Hal drove $2\frac{1}{2}$ hours at a constant speed of 66 mph on the first leg of a trip and $3\frac{3}{4}$ at a constant speed of 56 mph on the second leg of the trip. Find the average speed he drove over the entire trip.

    (A) 63 mph   **(B)** $59\frac{1}{2}$ mph   **(C)** $61\frac{1}{4}$ mph   **(D)** 61 mph   **(E)** 60 mph

Section 6: Exponents and Order of Operations

Exponents

Recall from Chapter 1 that expressions containing repeated multiplication of the same factor can be written in a condensed form called *exponent form*. For instance, the expression $\frac{3}{4} \cdot \frac{3}{4}$ can be written in exponent form as $\left(\frac{3}{4}\right)^2$, which is read as "three-fourths to the second power." In the expression, $\frac{3}{4}$ is called the *base* and 2 is called the *exponent*. The exponent indicates the number of times the base appears as a factor. For example, $\left(\frac{1}{2}\right)^5$ means to multiply 5 factors of $\frac{1}{2}$. That is,

$$\left(\frac{1}{2}\right)^5 = \underbrace{\frac{1}{2} \cdot \frac{1}{2} \cdot \frac{1}{2} \cdot \frac{1}{2} \cdot \frac{1}{2}}_{\text{5 factors of } \frac{1}{2}}$$

To evaluate expressions containing exponents, first rewrite the expression in terms of multiplication and then calculate the product.

EXAMPLE 1 Evaluate.

(a) $\left(\frac{2}{3}\right)^3$ (b) $\left(-\frac{4}{5}\right)^2$ (c) $\left(1\frac{1}{2}\right)^4$

SOLUTION

(a) $\left(\frac{2}{3}\right)^3 = \frac{2}{3} \cdot \frac{2}{3} \cdot \frac{2}{3} = \frac{2 \cdot 2 \cdot 2}{3 \cdot 3 \cdot 3} = \frac{8}{27}$

(b) $\left(-\frac{4}{5}\right)^2 = -\frac{4}{5} \cdot \left(-\frac{4}{5}\right) = \frac{4 \cdot 4}{5 \cdot 5} = \frac{16}{25}$

(c) $\left(1\frac{1}{2}\right)^4 = 1\frac{1}{2} \cdot 1\frac{1}{2} \cdot 1\frac{1}{2} \cdot 1\frac{1}{2} = \frac{3}{2} \cdot \frac{3}{2} \cdot \frac{3}{2} \cdot \frac{3}{2} = \frac{3 \cdot 3 \cdot 3 \cdot 3}{2 \cdot 2 \cdot 2 \cdot 2} = \frac{81}{16}$

CHAPTER 2 FRACTIONS PREALGEBRA REVIEW

Order of Operations

The *order of operations agreement* discussed in Chapter 1 applies to evaluating expressions containing fractions.

Order of Operations
1. Do all calculations within grouping symbols, like ( ), [ ], { }, and | |, starting with the innermost grouping symbol.
2. Evaluate all expressions containing exponents.
3. Do all multiplications and divisions as they appear from left to right.
4. Do all additions and subtractions as they appear from left to right.

EXAMPLE 2 Evaluate.

$$\text{(a)} \quad 4 - \frac{1}{3} \cdot \frac{3}{4} \qquad \text{(b)} \quad \frac{1}{3} \div \frac{6}{7} + \frac{1}{2} \cdot \frac{5}{9}$$

SOLUTION

$$\text{(a)} \quad 4 - \frac{1}{3} \cdot \frac{3}{4} = 4 - \frac{1 \cdot \cancel{3}}{\cancel{3} \cdot 4} = 4 - \frac{1}{4}$$

$$= \frac{4}{1} - \frac{1}{4}$$

$$= \frac{16}{4} - \frac{1}{4} = \frac{15}{4} = 3\frac{3}{4}$$

$$\text{(b)} \quad \frac{1}{3} \div \frac{6}{7} + \frac{1}{2} \cdot \frac{5}{9} = \frac{1}{3} \cdot \frac{7}{6} + \frac{1}{2} \cdot \frac{5}{9}$$

$$= \frac{7}{18} + \frac{5}{18} = \frac{12}{18}$$

$$= \frac{\cancel{6} \cdot 2}{\cancel{6} \cdot 3} = \frac{2}{3}$$

PREALGEBRA REVIEW CHAPTER 2   FRACTIONS

EXAMPLE 3   Calculate: $\left(\dfrac{3}{4}\right)^2 \div \dfrac{3}{8} + \dfrac{5}{12}$

SOLUTION

$$\left(\dfrac{3}{4}\right)^2 \div \dfrac{3}{8} + \dfrac{5}{12} = \dfrac{9}{16} \div \dfrac{3}{8} + \dfrac{5}{12}$$

$$= \dfrac{9}{16} \cdot \dfrac{8}{3} + \dfrac{5}{12}$$

$$= \dfrac{9 \cdot 8}{16 \cdot 3} + \dfrac{5}{12}$$

$$= \dfrac{\cancel{3} \cdot 3 \cdot \cancel{2} \cdot \cancel{2} \cdot \cancel{2}}{\cancel{2} \cdot \cancel{2} \cdot \cancel{2} \cdot 2 \cdot \cancel{3}} + \dfrac{5}{12}$$

$$= \dfrac{3}{2} + \dfrac{5}{12}$$

$$= \dfrac{18}{12} + \dfrac{5}{12}$$

$$= \dfrac{23}{12} = 1\dfrac{11}{12}$$

EXAMPLE 4   Evaluate: $\left(\dfrac{5}{9} - 1\dfrac{1}{3}\right) \cdot \dfrac{1}{4} + \dfrac{7}{8}$

SOLUTION

$$\left(\dfrac{5}{9} - 1\dfrac{1}{3}\right) \cdot \dfrac{1}{4} + \dfrac{7}{8} = \left(\dfrac{5}{9} - \dfrac{4}{3}\right) \cdot \dfrac{1}{4} + \dfrac{7}{8}$$

$$= \left(\dfrac{5}{9} - \dfrac{12}{9}\right) \cdot \dfrac{1}{4} + \dfrac{7}{8}$$

$$= \left(-\dfrac{7}{9}\right) \cdot \dfrac{1}{4} + \dfrac{7}{8}$$

$$= -\dfrac{7}{36} + \dfrac{7}{8}$$

$$= -\dfrac{14}{72} + \dfrac{63}{72}$$

$$= \dfrac{49}{72}$$

CHAPTER 2 FRACTIONS

Section 6: Exercises

Evaluate.

1. $\left(\dfrac{1}{2}\right)^5$
2. $\left(\dfrac{1}{3}\right)^3$
3. $\left(-\dfrac{3}{4}\right)^3$
4. $\left(-\dfrac{7}{9}\right)^2$
5. $\left(2\dfrac{2}{3}\right)^2$
6. $\left(-2\dfrac{1}{2}\right)^3$

Calculate. Write answers in lowest terms.

7. $\dfrac{3}{5} - \dfrac{2}{3} + \dfrac{8}{15}$
8. $\dfrac{2}{3} - \dfrac{8}{9} - \dfrac{4}{9}$

9. $3 \div \dfrac{1}{2} - \dfrac{3}{4}$
10. $\dfrac{1}{3} + \dfrac{1}{6} \cdot \dfrac{1}{5}$

11. $12 \div \dfrac{3}{4}(8)$
12. $\dfrac{5}{9} \div \dfrac{2}{3} \div 10$

13. $1\dfrac{3}{4} - \dfrac{3}{5} \cdot \dfrac{1}{2} + \dfrac{9}{10}$
14. $\dfrac{1}{6} + \dfrac{2}{5} \div \dfrac{8}{15} - 2$

15. $\dfrac{5}{8} - \dfrac{1}{3} + \left(-\dfrac{3}{4}\right)^2$
16. $\left(\dfrac{2}{3}\right)^2 - 4\left(-\dfrac{9}{16}\right)$

17. $\left(\dfrac{2}{3} - \dfrac{5}{6}\right) + \left(\dfrac{1}{2} - \dfrac{1}{8}\right)$
18. $\dfrac{3}{10} \div \left(1\dfrac{1}{5} + \dfrac{1}{4}\right) \times \dfrac{3}{8} - \dfrac{1}{9}$

19. $2\dfrac{1}{2} - \left(1\dfrac{2}{3}\right)^2 \div \dfrac{2}{9} + 3 \times \dfrac{1}{6}$
20. $4 - 3\dfrac{3}{4} \div \dfrac{5}{8} + \left(\dfrac{1}{2}\right)^3 \cdot \dfrac{4}{5}$

PREALGEBRA REVIEW

CHAPTER 2   FRACTIONS

Section 7: Complex Fractions

A **complex fraction** is a fraction in which the numerator, denominator, or both contains one or more fractions. For instance,

$$\frac{\frac{1}{3}}{\frac{1}{4}} \quad \leftarrow \text{Main fraction bar}$$

is a complex fraction. The main fraction bar in a complex fraction can be interpreted as division.

To simplify a complex fraction, rewrite the expression as a division problem.

EXAMPLE 1   Simplify: $\dfrac{\frac{1}{2}}{\frac{7}{8}}$

SOLUTION

$$\frac{\frac{1}{2}}{\frac{7}{8}} = \frac{1}{2} \div \frac{7}{8}$$

$$= \frac{1}{2} \cdot \frac{8}{7} = \frac{1 \cdot 8}{2 \cdot 7} = \frac{1 \cdot \cancel{2} \cdot 2 \cdot 2}{\cancel{2} \cdot 7} = \frac{4}{7}$$

EXAMPLE 2   Simplify: $\dfrac{1\frac{3}{5}}{\frac{8}{15}}$

SOLUTION

$$\frac{1\frac{3}{5}}{\frac{8}{15}} = 1\frac{3}{5} \div \frac{8}{15}$$

$$= \frac{8}{5} \div \frac{8}{15} = \frac{8}{5} \cdot \frac{15}{8} = \frac{8 \cdot 15}{5 \cdot 8} = \frac{\cancel{2} \cdot \cancel{2} \cdot \cancel{2} \cdot 3 \cdot \cancel{5}}{\cancel{5} \cdot \cancel{2} \cdot \cancel{2} \cdot \cancel{2}} = \frac{3}{1} = 3$$

CHAPTER 2 FRACTIONS

EXAMPLE 3 Simplify: $\dfrac{\frac{10}{13}}{-5}$

SOLUTION

$$\frac{\frac{10}{13}}{-5} = \frac{10}{13} \div (-5) = \frac{10}{13} \div \left(-\frac{5}{1}\right) = \frac{10}{13} \cdot \left(-\frac{1}{5}\right) = -\frac{10 \cdot 1}{13 \cdot 5} = -\frac{2 \cdot \cancel{5} \cdot 1}{13 \cdot \cancel{5}} = -\frac{2}{13}$$

EXAMPLE 4 Simplify.

(a) $\dfrac{2}{\frac{1}{3}+\frac{5}{6}}$ (b) $\dfrac{\frac{1}{2}-\frac{2}{5}}{\frac{3}{4}+\frac{1}{6}}$

SOLUTION First simplify the expressions in the numerator and denominator to obtain single fractions in the numerator and denominator.

(a) $\dfrac{2}{\frac{1}{3}+\frac{5}{6}} = \dfrac{2}{\frac{2}{6}+\frac{5}{6}}$

$= \dfrac{2}{\frac{7}{6}}$

$= 2 \div \dfrac{7}{6} = \dfrac{2}{1} \cdot \dfrac{6}{7} = \dfrac{12}{7} = 1\dfrac{5}{7}$

(b) $\dfrac{\frac{1}{2}-\frac{2}{5}}{\frac{3}{4}+\frac{1}{6}} = \dfrac{\frac{5}{10}-\frac{4}{10}}{\frac{9}{12}+\frac{2}{12}}$

$= \dfrac{\frac{1}{10}}{\frac{11}{12}}$

$= \dfrac{1}{10} \div \dfrac{11}{12} = \dfrac{1}{10} \cdot \dfrac{12}{11} = \dfrac{1 \cdot 12}{10 \cdot 11} = \dfrac{1 \cdot \cancel{2} \cdot 2 \cdot 3}{\cancel{2} \cdot 5 \cdot 11} = \dfrac{6}{55}$

Section 7: Exercises

Simplify.

1. $\dfrac{\frac{1}{9}}{\frac{5}{6}}$

2. $\dfrac{\frac{3}{4}}{\frac{3}{8}}$

3. $\dfrac{\frac{2}{4}}{5}$

4. $\dfrac{\frac{7}{12}}{9}$

5. $\dfrac{\frac{9}{10}}{\frac{18}{25}}$

6. $\dfrac{\frac{7}{16}}{\frac{21}{32}}$

7. $\dfrac{-\frac{20}{27}}{\frac{35}{36}}$

8. $\dfrac{-\frac{20}{21}}{\frac{-7}{12}}$

9. $\dfrac{-3}{\frac{9}{16}}$

10. $\dfrac{\frac{-12}{25}}{8}$

11. $\dfrac{\frac{4}{9}}{3\frac{1}{3}}$

12. $\dfrac{2\frac{1}{2}}{\frac{3}{5}}$

13. $\dfrac{1\frac{5}{6}}{2\frac{3}{8}}$

14. $\dfrac{4\frac{1}{5}}{3\frac{4}{7}}$

CHAPTER 2 FRACTIONS

15. $\dfrac{5\frac{1}{4}}{7}$

16. $\dfrac{9}{6\frac{3}{10}}$

17. $\dfrac{\frac{1}{3}}{\frac{1}{2}+\frac{1}{4}}$

18. $\dfrac{\frac{3}{5}-\frac{1}{10}}{\frac{1}{6}}$

19. $\dfrac{4+\frac{1}{3}}{\frac{2}{3}-\frac{1}{8}}$

20. $\dfrac{\frac{3}{8}+\frac{3}{4}}{2-\frac{1}{6}}$

21. $\dfrac{\frac{5}{6}+\frac{5}{8}}{\frac{4}{9}+\frac{1}{2}}$

22. $\dfrac{\frac{3}{4}-\frac{2}{5}}{\frac{2}{3}-\frac{7}{12}}$

23. $\dfrac{3\frac{1}{8}}{\frac{8}{15}-\frac{5}{6}}$

24. $\dfrac{\frac{9}{16}+\frac{11}{12}}{-2\frac{5}{9}}$

Chapter 3   Decimals

Section 1: Introduction to Decimals

Fractions with denominators that are powers of 10, such as $\frac{3}{10}$ and $\frac{57}{1000}$, are called **decimal fractions.** These fractions can be written in **decimal notation.** For instance, $\frac{3}{10}$ written in decimal notation is 0.3. A number written in decimal notation has a whole number part and a fractional part separated by a dot called a **decimal point.** A fraction written in decimal notation is called a **decimal number.**

Decimal Fraction	Decimal Number
$\frac{39}{100}$	0.39
$\frac{723}{1000}$	0.723
$\frac{609}{100}$	6.09
$3\frac{7}{10}$	3.7

Like fractions, decimal numbers represent parts of wholes.

Place Value and Word Names

The places to the right of a decimal point in a decimal number are called **decimal places.** Just as each digit of a whole number has a place value, each digit in the fractional part of a decimal number has a place value. To determine the place values of the fractional part of a decimal number, the place value chart for whole numbers is extended. Note that the place values for the whole number part are powers of 10: 1, 10, 100, 1000, and so on. The place values for the fractional part of a decimal number are reciprocals of powers of ten: $\frac{1}{10}$, $\frac{1}{100}$, $\frac{1}{1000}$, and so on.

CHAPTER 3   DECIMALS   PREALGEBRA REVIEW

| Place Value Chart ||||||||||
Whole number part						Fractional part			
Ten-thousands	Thousands	Hundreds	Tens	Ones	.	Tenths	Hundredths	Thousandths	Ten-thousandths
10,000	1000	100	10	1		$\frac{1}{10}$	$\frac{1}{100}$	$\frac{1}{1000}$	$\frac{1}{10,000}$

Note that a whole number has an unwritten decimal point to the right of the ones digit, with a 0 fractional part. For instance, 13 can be written as 13.0, 13.00, 13.000, and so on.

EXAMPLE 1   Identify the place value of each digit in the decimal number 152.38.

SOLUTION

Hundreds	Tens	Ones	.	Tenths	Hundredths
1	5	2	.	3	8

The decimal number 27.365 can be written in *expanded form* as follows.

$$27.365 = 2 \text{ tens } + 7 \text{ ones } + 3 \text{ tenths } + 6 \text{ hundredths } + 5 \text{ thousandths}$$

$$= 2(10) + 7(1) + 3\left(\frac{1}{10}\right) + 6\left(\frac{1}{100}\right) + 5\left(\frac{1}{1000}\right)$$

$$= 20 + 7 + \frac{3}{10} + \frac{6}{100} + \frac{5}{1000}$$

Using the least common denominator 1000 to combine the fractions gives

$$27.365 = 20 + 7 + \frac{300}{1000} + \frac{60}{1000} + \frac{5}{1000} = 27\frac{365}{1000}$$

Both 27.365 and $27\frac{365}{1000}$ are read as "twenty-seven and three hundred sixty-five thousandths."

Note that when the mixed number is written as a decimal number, the whole number part is to the left of the decimal point and the fractional part is to the right of the decimal point.

The word name for a decimal number is the same as the word name for the corresponding fraction or mixed number. Note that the decimal point is read "and." To write a word name from decimal notation, first, write a word name for the whole number part, next write "and" for the decimal point, and then, write a word name for the fractional part followed by the place value of the last digit. Note that if the whole number part is 0, then simply write the word name for the fractional part.

EXAMPLE 2 Write a word name for the decimal number.

(a) 643.23   (b) 0.0561

SOLUTION

(a) Six hundred forty-three and twenty-three hundredths
(b) Five hundred sixty-one thousandths

Writing Decimals as Fractions and Fractions as Decimals

Place value and the ability to read decimal numbers will help when writing decimals in fraction notation and fractions in decimal notation. Consider, for instance, the number thirty-nine hundredths, shown in both decimal notation and in fraction notation below. In fraction notation, 39 is the numerator of the fraction and the denominator is 100, which indicates the decimal place value hundredths. Note that the place value indicates a denominator that contains a power of 10 with the same number of zeros as the number of decimal places in the decimal number.

Decimal notation

0.39
2 decimal places

Fraction notation

$$\frac{39}{100}$$
2 zeros

In general, to write a decimal number in fraction notation, starting with the first nonzero digit, write all the digits in the number over a denominator that is a power of 10 that has the same number of zeros as the number of decimal places in the original decimal number.

When writing a decimal in fraction notation, do not simplify the fraction.

EXAMPLE 3 Write the decimal number in fraction notation.

(a) 0.631   (b) 0.07   (c) 0.0055

SOLUTION

(a) The decimal number has 3 decimal places. Write the digits in the number over 1000, which is a power of ten with 3 zeros.

$$\frac{631}{1000}$$

Note that the number in decimal notation and in fraction notation is read "six hundred thirty-one thousandths."

(b) $\frac{7}{100}$

(c) $\frac{55}{10,000}$

Decimal numbers greater than 1 can be written as a fraction or as a mixed number.

Example 4 Write 2.7 as a fraction and as a mixed number.

Solution To write the decimal number as a fraction, follow the given procedure.

$$2.7 = \frac{27}{10}$$

To write the decimal number as a mixed number, simply write the whole number part and express the decimal part as a fraction.

$$2.7 = 2\frac{7}{10}$$

Note that $\frac{27}{10}$ is an improper fraction that is equivalent to the mixed number $2\frac{7}{10}$.

To write a fraction as a decimal, the process is reversed. Write the digits in the numerator and starting from the rightmost digit and moving left, insert the decimal point so that the number has the same number of decimal places as the number of zeros as the power of 10 in the denominator.

If there are more zeros in the denominator than digits in the numerator, it will be necessary to insert zeros so that the decimal has the correct number of decimal places.

EXAMPLE 5 Write the fraction or mixed number in decimal notation.

(a) $\dfrac{9}{10}$ (b) $\dfrac{21}{10,000}$ (c) $2\dfrac{37}{100}$ (d) $\dfrac{2499}{100}$

SOLUTION

(a) 0.9

(b) 0.0021
Note that 2 zeros were inserted so that the decimal has the correct number of decimal places.

(c) 2.37

(d) 24.99

Just as with fractions, decimal numbers may be positive or negative. On the number line, positive decimal numbers are to the right of zero and negative decimal numbers are to the left of zero.

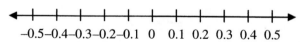

Negative decimal numbers are written in fraction notation in the same way that positive decimal numbers are written in fraction notation and vice versa.

EXAMPLE 6 Write –0.24 in fraction notation.

SOLUTION $-\dfrac{24}{100}$

EXAMPLE 7 Write $-\dfrac{17}{1000}$ in decimal notation.

SOLUTION –0.017

CHAPTER 3   DECIMALS   PREALGEBRA REVIEW

Section 1: Exercises

Identify the digit in each place value for the number 1540.65972.

1. Hundredths
2. Tens
3. Ten-thousandths
4. Thousandths
5. Thousands
6. Hundred-thousandths

Write the number in expanded notation.

7. 0.379
8. 1.2
9. 94.75
10. 4. 02614

Write in standard form.

11. 8 tenths + 2 hundredths

12. 9 ones + 0 tenths + 7 hundredths + 6 thousandths

13. 4 hundreds + 6 tens + 1 one + 9 tenths + 0 hundredths + 8 thousandths + 3 ten-thousandths

14. 1 thousand + 5 hundreds + 0 tens + 0 ones + 9 tenths

Write a word name.

15. 7.15
16. 0.431
17. 250.5
18. 23.0872

Write the decimal number as a fraction and, if possible, a mixed number.

19. 0.4
20. 0.567
21. 0.01
22. 0.0023
23. −0.233
24. −0.0009
25. 1.65
26. 21.6
27. −9.425
28. −142.99
29. 5.02
30. 16.003

Write the decimal fraction in decimal notation.

31. $\dfrac{13}{100}$ 32. $\dfrac{2}{10}$ 33. $\dfrac{451}{10,000}$ 34. $\dfrac{8}{1000}$

35. $-\dfrac{45}{100}$ 36. $-\dfrac{407}{1000}$ 37. $3\dfrac{7}{100}$ 38. $10\dfrac{1}{10}$

39. $-12\dfrac{67}{1000}$ 40. $\dfrac{3459}{100}$ 41. $-\dfrac{10,001}{10,000}$ 42. $\dfrac{2008}{10}$

Choose the correct answer.

43. Which number is not equivalent to one and thirty-two thousandths?

   (A) 1.032   (B) $1\dfrac{32}{1000}$ (C) $\dfrac{132}{1000}$ (D) 1.32   (E) $1(1)+32\left(\dfrac{1}{1000}\right)$

44. Which number is equivalent to $3(10)+9(1)+1\left(\tfrac{1}{10}\right)+7\left(\tfrac{1}{100}\right)+5\left(\tfrac{1}{1000}\right)$?

   (A) 3.91175   (B) 39.175   (C) 391.075   (D) 309.175   (E) 39.0175

45. What decimal is represented by the shaded portion in the following figure?

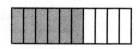

   (A) 0.6   (B) $\dfrac{6}{10}$ (C) 0.06   (D) 0.5   (E) 0.4

46. Which number is not equivalent to 5?

   (A) 5.0   (B) $\dfrac{500}{100}$ (C) $\dfrac{50}{10}$ (D) $\dfrac{5}{10}$ (E) 5.000

CHAPTER 3   DECIMALS   PREALGEBRA REVIEW

Section 2: Comparing and Rounding Decimals

Comparing Decimals

Consider the decimal numbers 0.3 and 0.25. Note that 0.3 is equivalent to 0.30 because $\frac{3}{10}$ is equivalent to $\frac{30}{100}$. Since $0.25 = \frac{25}{100}$, it follows that $\frac{30}{100} > \frac{25}{100}$ and $0.3 > 0.25$.

Procedure for Comparing Two Decimal Numbers:
Write the decimal numbers in fraction notation with a common denominator. Then, compare the fractions.

EXAMPLE 1   Write > or < between each pair of numbers to make a true statement.

(a) 0.79 _____ 0.92
(b) 1.475 _____ 1.471
(c) 0.679 _____ 0.6798

SOLUTION

(a) First, write each decimal in fraction notation: $0.79 = \frac{79}{100}$ and $0.92 = \frac{92}{100}$

The fractions have a common denominator of 100. Next, compare the fractions: $\frac{79}{100} < \frac{92}{100}$. Since $\frac{79}{100} < \frac{92}{100}$, $0.79 < 0.92$

(b) $1.475 = \frac{1475}{1000}$ and $1.471 = \frac{1471}{1000}$

The fractions have a common denominator of 1000. Compare the fractions: $\frac{1475}{1000} > \frac{1471}{1000}$. So, $1.475 > 1.471$

(c) $0.679 = \frac{679}{1000}$ and $0.6798 = \frac{6798}{10,000}$

Multiply the numerator and denominator of $\frac{679}{1000}$ by 10 so that the fractions have a common denominator of 10,000: $\frac{679 \cdot 10}{1000 \cdot 10} = \frac{6790}{10,000}$.

Compare the fractions: $\frac{6790}{10,000} < \frac{6798}{10,000}$. So, $0.679 < 0.6798$

116

PREALGEBRA REVIEW   CHAPTER 3   DECIMALS

EXAMPLE 2   List the numbers 0.436, 0.00436, 4.36, 0.4036, 0.0436 in order from smallest to largest.

SOLUTION   Write each decimal number in fraction notation with the common denominator 100,000 and then compare the fractions.

$$\frac{436}{100,000} < \frac{4360}{100,000} < \frac{40,360}{100,000} < \frac{43,600}{100,000} < \frac{436,000}{100,000}$$

So, 0.00436, 0.0436, 0.4036, 0.436, 4.36

Absolute Value

Recall that the absolute value of a number is its distance from 0 on the number line. For instance, $|-0.2| = 0.2$ because it is 0.2 units from 0 on the number line.

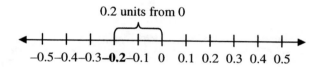

EXAMPLE 3   Determine the absolute value of the number.

(a) $|1.45|$ (b) $|-13.001|$ (c) $|-0.2355|$ (d) $|9.1|$

SOLUTION

(a) 1.45   (b) 13.001   (c) 0.2355   (d) 9.1

EXAMPLE 4   Insert > or < between each pair of numbers to make a true statement.

(a) $|3.6|$ ____ $|-4.7|$
(b) -0.919 ____ $|-0.919|$
(c) $|-8.235|$ ____ $|-8.23|$

SOLUTION

(a) $|3.6| = 3.6$ and $|-4.7| = 4.7$
    $3.6 < 4.7$, so $|3.6| < |-4.7|$
(b) $|-0.919| = 0.919$
    $-0.919 < 0.919$, so $-0.919 < |-0.919|$
(c) $|-8.235| > |-8.23|$

CHAPTER 3   DECIMALS

Rounding Decimals

Decimal approximations are often needed when working with decimal numbers. To approximate a decimal number, the number is **rounded** to a specific place value. For instance, the number 2.47865 rounded to the nearest tenth is 2.5. So, 2.47865 is approximately equal to 2.5.

> **Procedure for Rounding a Decimal Number:**
> Locate the digit in the place value to which the number is to be rounded. Then, look at the next digit to the right. If the digit to the right is greater than or equal to 5, add one to the original digit. If the digit to the right less than 5, keep the original digit the same. Drop all the digits to the right of the place value to which the number is rounded.

EXAMPLE 3   Round 2.795321 to the nearest

   (a) tenth   **(b)** ten-thousandth   **(c)** hundredth

SOLUTION
(a) The digit in the tenths place is 7. The next digit to the right is 9. Since 9 is greater than 5, add one to 7 and drop all the digits to the right of the tenths place. So, 2.795321 rounded to the nearest tenth is 2.8.
(b) The digit in the ten-thousandths place is 3. The next digit to the right is 2. Since 2 is less than 5, keep the 3 the same and drop all the digits to the right of the ten-thousandths place. So, 2.795321 rounded to the nearest ten-thousandth is 2.7953.
(c) The digit in the hundredths place is 9. The next digit to the right is 5. Since 5 is equal to 5, add one to the original digit and drop all the digits to the right of the hundredths place. So, 2.795321 rounded to the nearest hundredth is 2.80. (Note that when 1 is added to 9, the sum is 10. The digit in the hundredths place becomes 0 and the 1 is carried to the tenths place.)

PREALGEBRA REVIEW   CHAPTER 3   DECIMALS

Section 2: Exercises

Write > or < between each pair of decimal numbers to make a true statement.

1. 0.6 ____ 0.1   2. 1.04 ____ 1.09   3. 0.372 ____ 0.377

4. 2.9 ____ 2.97   5. 0.023 ____ 0.0023   6. 1.2456 ____ 1.245

List the decimal numbers in order from smallest to largest.

7. 0.598, 0.5098, 0.0589, 0.55908, 0.5

8. 4.113, 41.13, 4.13, 14.31, 4.11

List the decimal numbers in order from largest to smallest.

9. 0.099, 0.99, 0.009, 0.909, 0.09

10. 60.2, 6.22, 0.602, 0.062, 62.2

Determine the absolute value of the number.

11. $|-0.63|$ 12. $|0.7021|$ 13. $|9.093|$ 14. $|-5.5|$

Insert >, < or = between each pair of numbers to make a true statement.

15. $|-6.13|$ ____ $|-5.2|$ 16. 0.8 ____ $|-0.8|$ 17. $|-0.28|$ ____ $|0.289|$

18. $\left|-\frac{3}{100}\right|$ ____ $|0.3|$ 19. $|-1.05|$ ____ $\left|-1\frac{5}{1000}\right|$ 20. $\left|-\frac{143}{10}\right|$ ____ -14.3

Round the decimal number to the indicated place value.

21. 2.7891 to the nearest hundredth   22. 0.95 to the nearest tenth

23. 1.00606 to the nearest thousandth   24. 4.037 to the nearest tenth

25. 7.998 to the nearest hundredth   26. 0.499775 to the nearest thousandth

27. 5.864 to the nearest tenth

28. 297.88 to the nearest hundred

Choose the correct answer.

29. Which number is not equivalent to 0.5?

    (A) 0.500   **(B)** $\dfrac{5}{10}$ **(C)** 0.05   **(D)** 0.50   **(E)** $\dfrac{50}{100}$

30. Which number is equivalent to 0.003?

    (A) $\dfrac{3}{10}$ **(B)** $\dfrac{30}{10,000}$ **(C)** $\dfrac{30}{100}$ **(D)** $\dfrac{3}{100}$ **(E)** $\dfrac{3}{10,000}$

31. Which number is greater than four and twenty-three thousandths?

    (A) 4.23   **(B)** 4.023   **(C)** 0.423   **(D)** 4.0023   **(E)** 0.0424

32. Which number is less than two and seven hundredths?

    (A) 2.7   **(B)** 0.27   **(C)** 2.07   **(D)** 27.7   **(E)** 2.27

33. Which number is 41.67959 rounded to the nearest thousandth?

    (A) 41.679   **(B)** 41.6796   **(C)** 41.7   **(D)** 41.68   **(E)** 41.680

34. The tax on a purchase of $23.56 is calculated to be $1.1785. Round this amount to the nearest cent.

    (A) $1.18   **(B)** $1.2   **(C)** $1.179   **(D)** $1.17   **(E)** $1.178

Section 3: Addition and Subtraction of Decimals

Addition of Decimals

Addition of decimal numbers is similar to addition of whole numbers.

To add two or more decimal numbers, write the numbers vertically with the decimal points and corresponding place values lined up. If necessary, insert zeros so that each addend has the same number of decimal places. Then add the digits in each column starting from the right, "carrying" if necessary. The decimal point in the sum should be lined up with the decimal points in the addends.

EXAMPLE 1   Add: 42.649 + 8.37

SOLUTION

$$\begin{array}{r} \overset{11}{4}2.649 \\ +8.370 \\ \hline 51.019 \end{array}$$ ← Insert a 0 so that the number of decimal places is the same.

EXAMPLE 2   Add: 1 + 0.69873 + 0.875

SOLUTION

$$\begin{array}{r} \overset{111}{1}.00000 \\ 0.69873 \\ +0.87500 \\ \hline 2.57373 \end{array}$$

Subtraction of Decimals

As with addition, subtraction of decimal numbers is similar to subtraction of whole numbers.

To subtract two decimal numbers, write the numbers vertically with the decimal points and corresponding place values lined up. If necessary, insert zeros so that the minuend and subtrahend have the same number of decimal places. Then subtract the digits in each column starting from the right, "borrowing" if necessary. The decimal point in the difference should be lined up with the decimal points in the minuend and subtrahend.

CHAPTER 3 DECIMALS   PREALGEBRA REVIEW

EXAMPLE 3   Subtract: 13.91 − 9.712

SOLUTION

$$\begin{array}{r} 13\overset{\,8\,10\,10}{.\cancel{9}\cancel{1}\cancel{0}} \\ -\ \ 9.7\,1\,2 \\ \hline 4.1\,9\,8 \end{array}$$

Addition and Subtraction of Signed Decimals

The procedures for adding and subtracting integers work for adding and subtracting positive and negative decimal numbers.

Procedure for Adding Signed Numbers:
- *Addends have the same sign* – Find the absolute values of the numbers and add them. Use the common sign as the sign of the sum. If both addends are positive, the sum is positive. If both addends are negative, the sum is negative.
- *Addends have different signs* – Find the absolute values of the numbers and subtract the smaller absolute value from the larger absolute value. Use the sign of the number with the larger absolute value as the sign of the sum.

EXAMPLE 4   Add.

   (a) 2.74 + (−5.8)   (b) −8.457 + (−13)

SOLUTION

(a) Find the absolute values of the numbers.

$|2.74| = 2.74$ and $|-5.8| = 5.8$

Subtract: 5.8 − 2.74 = 3.06
Since −5.8 has the larger absolute value and is negative, the sum is negative.

2.74 + (−5.8) = −3.06

(b) $|-8.457| = 8.457$ and $|-13| = 13$

Add: 8.457 + 13 = 21.457
Since both addends are negative, the sum is negative.

−8.457 + (−13) = −21.457

PREALGEBRA REVIEW

CHAPTER 3 DECIMALS

Recall that when subtracting integers, the key was to change the subtraction problem to a related addition problem.

Procedure for Subtracting Signed Numbers:
To subtract $a - b$, add the opposite, or additive inverse of b to a. That is, $a - b = a + (-b)$.

EXAMPLE 5   Subtract.

(a) $17.23 - 25.154$ (b) $-0.8209 - (-3.77)$

SOLUTION

(a) $17.23 - 25.154 = 17.23 + (-25.154)$
$= -7.924$

(b) $-0.8209 - (-3.77) = -0.8209 + 3.77$
$= 2.9491$

Solving Applied Problems

EXAMPLE 6   Nina bought a gallon of milk for $3.29, a loaf of bread for $1.69 and a pack of gum for $0.89 at a convenience store. What was her total bill for the three items?

SOLUTION   The price of each item Nina purchased is given. To find the total bill, add.

Price of milk + price of bread + Price of gum   → $3.29 + $1.69 + $0.89

$3.29 + 1.69 + 0.89 = 5.87$

Nina's total bill was $5.87.

CHAPTER 3   DECIMALS   PREALGEBRA REVIEW

EXAMPLE 7   After taking aspirin, Ciara's fever dropped from 102.7° to 99.1°. How many degrees did her temperature drop?

SOLUTION   Ciara's temperatures before she took aspirin and after she took aspirin are given. To find how many degrees her temperature dropped, subtract.

Temperature before aspirin − Temperature after aspirin   →   102.7° − 99.1°

102.7 − 99.1 = 3.6

Her temperature dropped 3.6°.

EXAMPLE 8   A contractor has 68 feet of crown molding to install in a rectangular dining room measuring 11.2 feet by 8.7 feet. How many feet of crown molding will he have left after he installs it in the dining room?

SOLUTION   The amount of crown molding and the dimensions of the dining room are given. To find the amount of crown molding left after it is installed, first find the amount needed in the dining room. To find this amount, calculate the perimeter of the dining room. The perimeter of a rectangle is the sum of the measures of the four sides. In this problem, it is helpful to draw and label a picture.

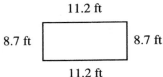

The perimeter of the rectangular dining room is:

11.2 + 8.7 + 11.2 + 8.7 = 39.8

Since the perimeter is 39.8 ft, the contractor will need 39.8 feet of crown molding for the dining room. To find the amount of crown molding left after the contractor installs it in the dining room, subtract.

Total amount of   −   Amount used in   →   68 ft − 39.8 ft
crown molding   dining room

68 − 39.8 = 28.2

The contractor will have 28.2 ft of crown molding left.

PREALGEBRA REVIEW   CHAPTER 3   DECIMALS

Section 3: Exercises

Add.

1. 3.12 + 9.83
2. 0.75 + 0.25
3. 21.042 + 7.3
4. 0.679 + 3.77299
5. 0.14 + 0.926 + 2
6. 4 + 0.7001 + 0.008
7. 43.206 + 18 + 5.0092
8. 0.5 + 88.6807 + 49

Subtract.

9. 2.31 − 1.79
10. 16.097 − 9.458
11. 0.8522 − 0.749
12. 9.002 − 0.70055
13. 1 − 0.9999
14. 10 − 0.01
15. 54.6 − 32.084
16. 81.9 − 65.91

Add or Subtract.

17. 5.33 + (−1.65)
18. −0.628 + 0.254
19. 8.4 − 9.07
20. 13.573 − (−10.01)
21. −24.24 + (−18.42)
22. −11.1 − 12.1
23. −7 − (−6.99)
24. −31.885 + 29

Choose the correct answer.

25. Subtract 5.7 from 3.03.

    (A) 2.04   (B) −2.67   (C) 2.46   (D) 2.67   (E) −2.73

CHAPTER 3   DECIMALS

26. How much more than 16.085 is 21?

(A) 37.085 (B) 5.015 (C) 4.15 (D) 4.015 (E) 4.915

27. What is the sum 2.976 and 8.64 rounded to the nearest tenth?

(A) 11.0 (B) 11.3 (C) 11.6 (D) 11.5 (E) 3.8

28. What is the sum of −0.825 and its opposite?

(A) 0 (B) 0.825 (C) 0.175 (D) −0.825 (E) 1

29. What is the difference between $\frac{19}{100}$ and 1.1?

(A) −0.91 (B) 1.29 (C) 20.1 (D) 0.8 (E) −1.09

30. What is the sum of 2.04 and $3\frac{7}{1000}$?

(A) 5.74 (B) 5.11 (C) 5.47 (D) 5.04 (E) 5.047

31. Lee pays $8.39 including sales tax for a paperback book. If the price of the book was $7.99, then how much sales tax did he pay?

(A) $1.30 (B) $0.60 (C) $0.40 (D) $1.60 (E) $0.39

32. The total land area of New York City, NY is 301.5 square miles and the total land area of Boston, MA is 47.2 square miles. How many square miles larger than Boston is New York City?

(A) 263.3 (B) 348.7 (C) 264.3 (D) 261.3 (E) 254.3

33. Ana drove 2.8 miles from her apartment to school. After school she drove 5.9 miles to work and then 8.4 miles home. How many miles did Ana drive?

(A) 8.7 (B) 17.1 (C) 11.2 (D) 14.3 (E) 11.5

34. Paget has a balance of $523.84 in her checking account. What is her new balance after she deposits a check for $387.29?

(A) $911.13 (B) $136.55 (C) $801.03 (D) $901.13 (E) $811.13

PREALGEBRA REVIEW   CHAPTER 3   DECIMALS

35. It takes 365.256 days for Earth to orbit the Sun. It takes 686.98 days for Mars to orbit the Sun. How many more days does it take Mars to orbit the Sun?

   (A) 301.72   **(B)** 321.734   **(C)** 311.724   **(D)** 321.724   **(E)** 1052.236

36. Dave jogged 6.44 kilometers this morning. His friend Hal jogged with him and kept going another 2.41 kilometers. How many kilometers did Hal jog?

   (A) 6.44   **(B)** 15.29   **(C)** 2.41   **(D)** 8.85   **(E)** 4.03

37. Max bought a DVD that cost $22.99. The sales tax on the purchase was $1.15. If he gave the cashier $30, how much change did he receive?

   (A) $7.01   **(B)** $8.16   **(C)** $5.86   **(D)** $21.84   **(E)** $6.86

38. Jim's gross pay this week was $487.42. Federal tax of $71.31 and state tax of $22.97 were deducted from his gross pay. What was his pay after taxes?

   (A) $416.11   **(B)** $393.14   **(C)** $464.45   **(D)** $394.14   **(E)** $94.28

39. A rectangular gymnastics mat measures 3.66 meters by 1.83 meters. What is the perimeter of the mat?

   (A) 5.49 m   **(B)** 9.15 m   **(C)** 7.32   **(D)** 10.98   **(E)** 8.88

40. Julie has 87.25 feet of fencing to enclose her rectangular garden that measures 12.4 feet by 10.8 feet. How many feet of fencing will be left once she encloses her garden?

   (A) 40.85   **(B)** 46.4   **(C)** 64.05   **(D)** 41.15   **(E)** 53.25

CHAPTER 3   DECIMALS   PREALGEBRA REVIEW

Section 4: Multiplication and Division of Decimals

Multiplication of Decimals

The decimals 0.4 and 0.8 can be multiplied by multiplying the equivalent fraction forms of the decimal numbers and then converting back to decimal notation.

$$0.07 \times 0.9 = \frac{7}{100} \times \frac{9}{10} = \frac{63}{1000} = 0.063$$

Notice that the product has three decimal places, which is the sum of the number of decimal places in the factors.

$$\begin{array}{rl} 0.07 & \leftarrow \text{ 2 decimal places} \\ \times\ 0.9 & \leftarrow\ +1 \text{ decimal place} \\ \hline 0.063 & \leftarrow \text{ 3 decimal places} \end{array}$$

> **Procedure for Multiplying Two Decimal Numbers:**
> Disregard the decimal points and multiply the factors as though they were whole numbers. Starting from the rightmost digit in the product and moving left, place the decimal point so that the product has the same number of decimal places as the sum of the decimal places in the factors. Note that extra zeros may need to be inserted to the left of the product in order to get the correct number of decimal places.

Notice that, unlike addition and subtraction, it is not necessary to line up the decimal points when multiplying decimal numbers.

EXAMPLE 1   Multiply: 3.29×7.45

SOLUTION   Disregard the decimal points and multiply the factors as though they were whole numbers.

$$\begin{array}{r} 3.29 \\ \times\ 7.45 \\ \hline 1645 \\ 13160 \\ 230300 \\ \hline 245105 \end{array}$$

Starting from the rightmost digit in the product and moving left, place the decimal point so that the product has the same number of decimal places as the sum of the decimal places in the factors.

$$\begin{array}{rl} 3.29 & \leftarrow \quad 2 \text{ decimal places} \\ \times \ 7.45 & \leftarrow \ +2 \text{ decimal places} \\ \hline 1645 & \\ 13160 & \\ 230300 & \\ \hline 24.5105 & \leftarrow \quad 4 \text{ decimal places} \end{array}$$

EXAMPLE 2 Multiply: 0.0148×0.67

$$\begin{array}{rl} 0.0148 & \leftarrow \quad 4 \text{ decimal places} \\ \times \quad 0.67 & \leftarrow \ +2 \text{ decimal places} \\ \hline 1036 & \\ 8880 & \\ \hline 0.009916 & \leftarrow \quad 6 \text{ decimal places} \end{array}$$

Notice that two zeros had to be inserted in the product to get the correct number of decimal places.

Division of Decimals

In Chapter 1, division of whole numbers using a process called "long division" was discussed. Long division involved bringing down numbers in the dividend until all that was left was a remainder. The long division process can be used to divide decimal numbers.

Procedure for Dividing a Decimal Number by a Whole Number:
Place the decimal point in the quotient directly above the decimal point in the dividend. Divide as though dividing whole numbers. Continue dividing until the remainder is 0, until the digits in the quotient begin to repeat in a pattern, or until the quotient has the desired number of decimal places. Note that extra zeros may need to be inserted to the right of the last digit in the dividend in order to continue the long division process.

CHAPTER 3 DECIMALS

EXAMPLE 3 Divide: $2.67 \div 3$

SOLUTION
```
      0.89
   3)2.67
     24
     ──
      27
      27
      ──
       0
```

EXAMPLE 4 Divide 236.2 by 21. Round the quotient to the nearest thousandth.

SOLUTION Four decimal places are needed in order to round the quotient to the nearest thousandth. Note that three zeros must be inserted in the dividend to get the correct number of decimal places in the quotient.

```
        11.2476
   21)236.2000
       21
       ──
        36
        21
        ──
         52
         42
         ──
         100
          84
         ───
         160
         147
         ───
          130
          126
          ───
            4
```

Rounded to the nearest thousandth, the quotient is 11.248.

To divide a decimal number by a decimal number, first change the divisor to a whole number. To see how, consider the division problem $2.16 \div 0.9$. Since a fraction bar indicates division, the problem can be written as $\frac{2.16}{0.9}$. Recall that multiplying the numerator and denominator by the same nonzero number gives an equivalent fraction. If the numerator and the denominator of the fraction $\frac{2.16}{0.9}$ are multiplied by 10, the result is an equivalent fraction with a whole number divisor.

$$\frac{2.16}{0.9} = \frac{2.16 \times 10}{0.9 \times 10} = \frac{21.6}{9}$$

Notice that multiplying the numerator and denominator by 10, moves the decimal point one place to the right. So, $\frac{2.16}{0.9}$ is equivalent to $\frac{21.6}{9}$. That is,

$$0.9 \overline{)2.16} \quad \xrightarrow{\text{is equivalent to}} \quad 9 \overline{)21.6}$$

Procedure for Dividing by a Decimal Number:
Count the number of decimal places in the divisor and move the decimal point that many places to the right. Move the decimal point in the dividend the same number of places to the right, inserting zeros if necessary. Write the decimal point in the quotient directly above the decimal point in the dividend. Then divide.

EXAMPLE 5 Divide: $1.28 \div 0.4$

SOLUTION The divisor has 1 decimal place. Move the decimal point in the divisor and in the dividend 1 place to the right.

$$0.4 \overline{)1.28} \quad \rightarrow \quad 04. \overline{)12.8}$$

Write the decimal point in the quotient directly above the decimal point in the dividend and divide.

```
      3.2
   4)12.8
     12
     ---
      08
       8
       --
       0
```

CHAPTER 3   DECIMALS

EXAMPLE 6   Divide 36.9 by 12.15. Round to the nearest thousandth.

SOLUTION   Note that a 0 must be inserted in the dividend.

$$12.15 \overline{)36.9} \rightarrow 1215. \overline{)3690.}$$

In order to round to the nearest thousandth, the division must be carried out to four decimal places.

```
         3.0370
  1215)3690.0000
       3645
        4500
        3645
        8550
        8505
         450
```

Rounding to the nearest thousandth gives 3.037.

Multiplying and Dividing Signed Decimals

The sign rules for multiplying and dividing integers work for multiplying and dividing decimals.

> **Sign Rules for Multiplying (or Dividing) Two Signed Numbers:**
> - *Same sign* – If the two numbers have the same sign, then the product (or quotient) is positive.
> - *Different signs* – If the two numbers have different signs, then the product (or quotient) is negative.

PREALGEBRA REVIEW   CHAPTER 3   DECIMALS

EXAMPLE 7   Multiply: $(-4.8)(-5.23)$

SOLUTION   Multiply the decimals.

$$\begin{array}{r} 5.23 \\ \times 4.8 \\ \hline 4\,184 \\ 20\,920 \\ \hline 25.104 \end{array}$$

Since a negative number is multiplied by a negative number, the product is positive. So, $(-4.8)(-5.23) = 25.104$.

EXAMPLE 8   Divide: $\dfrac{-1.364}{0.08}$

SOLUTION   Divide the decimals.

$$0.08\overline{)1.364} \quad \rightarrow \quad 0008.\overline{)136.4}$$

$$\begin{array}{r} 17.05 \\ 8\overline{)136.40} \\ \underline{8} \\ 56 \\ \underline{56} \\ 0\,40 \\ \underline{40} \\ 0 \end{array}$$

Since a negative number is divided by a positive number, the quotient is negative. So, $\dfrac{-1.364}{0.08} = -17.05$

CHAPTER 3   DECIMALS   PREALGEBRA REVIEW

Multiplying and Dividing Decimals by Powers of 10

Recall that the powers of 10 are 10, 100, 1000, 10,000, and so on. Consider the following products of 2.658 multiplied by a power of 10.

$$
\begin{array}{r} 2.658 \\ \times \quad 10 \\ \hline 26.580 \end{array}
\qquad
\begin{array}{r} 2.658 \\ \times \quad 100 \\ \hline 265.800 \end{array}
\qquad
\begin{array}{r} 2.658 \\ \times \quad 1000 \\ \hline 2658.000 \end{array}
$$

Each product is larger than 2.658 and contains the digits 2658. Note that multiplying by a power of 10 moves the decimal point in the original number to the *right* the same number of places as the number of zeros in the power of 10.

In general, to multiply a decimal by a power of 10, count the number of zeros in the power of 10 and move the decimal point that many places to the right, inserting zeros if necessary.

EXAMPLE 9   Multiply.

(a) 100×3.2 (b) $0.14587(10,000)$

SOLUTION

(a) There are two zeros in the power of 10, so move the decimal point two places to the right. Note that a 0 must be inserted.

$$100 \times 3.2 = 320$$

(b) $0.14587(10,000) = 1458.7$

Now consider the following quotients of 12.9 divided by a power of 10.

$$
\begin{array}{r}
1.29 \\
10\overline{)12.90} \\
\underline{10} \\
2\,9 \\
\underline{2\,0} \\
90 \\
\underline{90} \\
0
\end{array}
\qquad
\begin{array}{r}
0.129 \\
100\overline{)12.900} \\
\underline{10\,0} \\
2\,90 \\
\underline{2\,00} \\
900 \\
\underline{900} \\
0
\end{array}
\qquad
\begin{array}{r}
0.0129 \\
1000\overline{)12.9000} \\
\underline{10\,00} \\
2\,900 \\
\underline{2\,000} \\
9000 \\
\underline{9000} \\
0
\end{array}
$$

Each quotient is smaller than 12.9 and contains the digits 129. Note that dividing by a power of 10 moves the decimal point in the original number to the *left* the same number of places as the number of zeros in the power of 10.

In general, to divide a decimal by a power of 10, count the number of zeros in the power of 10 and move the decimal point that many places to the left, inserting zeros if necessary.

EXAMPLE 10 Divide.

(a) $\dfrac{142.76}{1000}$ (b) $1.193 \div 100{,}000$

SOLUTION

(a) There are three zeros in the power of ten, so move the decimal point three places to the left.

$$\dfrac{142.76}{1000} = 0.14276$$

(b) $1.193 \div 100{,}000 = 0.00001193$

Order of Operations

EXAMPLE 11 Simplify: $5.7 - 0.64 \div 1.6$

SOLUTION

$$5.7 - 0.64 \div 1.6 = 5.7 - 0.4$$
$$= 5.3$$

EXAMPLE 12 Evaluate: $1.08 \div 0.9 + (0.5)^2 - 1$

SOLUTION

$$1.08 \div 0.9 + (0.5)^2 - 1 = 1.08 \div 0.9 + 0.25 - 1$$
$$= 1.2 + 0.25 - 1$$
$$= 0.45$$

CHAPTER 3   DECIMALS

Solving Applied Problems

EXAMPLE 13   Ana bought three t-shirts at a department store. If each t-shirt cost $12.99, what was the total amount she paid for the shirts?

SOLUTION   The number of t-shirts Ana bought and the price she paid for each t-shirt are given. To find the total amount she paid, multiply.

Number of t-shirts × Price of one t-shirt   →   3 × $12.99

3 × 12.99 = 38.97

Ana paid a total of $38.97 for the t-shirts.

EXAMPLE 14   The total cost of six admissions to an amusement park cost $239.70. What is the price of one admission?

SOLUTION   The number of admissions and the total cost are given. To find the price of one admission, divide.

Total cost of admissions ÷ Number of admissions   → $239.70 ÷ 6

239.70 ÷ 6 = 39.95

The price of one admission is $39.95.

EXAMPLE 15   Mike builds a 206.25 square foot rectangular patio in his yard. How much longer than the width is the length of the patio if the width is 12.5 feet?

SOLUTION   The area and width of the patio are given. To find how much longer the length is, first find the length of the patio by dividing.

Area of patio ÷ Width of patio   →   206.25 ft^2 ÷ 12.5 ft

206.25 ÷ 12.5 = 16.5

The length is 16.5 feet. To determine how much longer the length is, subtract.

Length of patio − Width of patio   →   16.5 ft − 12.5 ft

16.5 − 12.5 = 4

The length is 4 feet longer than the width.

PREALGEBRA REVIEW   CHAPTER 3   DECIMALS

Section 4: Exercises

Multiply.

1. 4.3×6.1
2. 3.9×2.8
3. $(0.7)(0.5)$
4. $(1.2)(0.9)$
5. $3.46 \cdot 0.15$
6. $0.36 \cdot 7.333$
7. $5.25(6.4109)$
8. $2.783(4.936)$
9. 0.024×0.02
10. 0.006×9.103
11. $16 \cdot 4.251$
12. $8.0665 \cdot 11$

Divide.

13. $26.24 \div 4$
14. $30.17 \div 7$
15. $\dfrac{252.81}{18}$
16. $\dfrac{24.9}{25}$
17. $0.9 \overline{)1.116}$
18. $0.891 \div 1.25$
19. $0.00576 \div 0.36$
20. $0.564 \overline{)2.397}$
21. $\dfrac{0.001071}{0.0238}$
22. $\dfrac{215.914}{8.9}$
23. $198.83 \div 16.85$
24. $16.4 \overline{)295.61}$

Divide. Round to the nearest thousandth.

25. $8.45 \div 12$
26. $2.36 \div 0.3$
27. $\dfrac{0.074}{0.928}$
28. $\dfrac{5}{0.61}$

Multiply or divide.

29. $(-0.6)(3.9)$
30. $-1.25 \cdot (-4.68)$
31. $-22 \div 5.5$
32. $1.98 \div (-0.88)$
33. 8.3×100
34. $12.56 \div 1000$

35. $3.05 \div 10$ **36.** $0.0036(1000)$ **37.** $\dfrac{-8}{-0.025}$

38. $(-0.0032)(-0.045)$ **39.** $-3.89 \div 10{,}000$ **40.** $100{,}000(-13.0978)$

Simplify.

41. $8.6 - 0.8 \times 5$ **42.** $9 + 0.6 - 3.8 - 1.2$

43. $(0.4)^3 + (0.7)^2$ **44.** $0.064 \div (8.7 - 5.5) \cdot 10^3$

45. $(0.3 + 0.9)^2 + 5 \times 1.2$ **46.** $0.008 \cdot (7.8 + 2.2) - 2.4 \cdot 0.01$

Choose the correct answer.

47. Carly drove for 2.5 hours at a constant speed of 63 miles per hour. How many miles did she drive?

    (A) 158   **(B)** 25.2   **(C)** 15.75   **(D)** 157.5   **(E)** 252

48. Ella pays $1.59 for a large cup of coffee at a coffee shop. How much does she spend on five large cups of coffee?

    (A) $7.95   **(B)** $6.36   **(C)** $0.795   **(D)** $8.00   **(E)** $3.18

49. Phil's gross monthly salary is $2704.25. What is his gross annual salary?

    (A) $27,042.50   **(B)** $37,859.50   **(C)** $32,415.00
    (D) $34,521.00   **(E)** $32,451.00

50. A phone company charges $0.07 per minute for long distance phone calls. If a customer's long distance phone bill was $10.78 this month, how many minutes of long distance calls was he charged for?

    (A) 1540   **(B)** 162   **(C)** 154   **(D)** 145   **(E)** 15.4

51. A dozen muffins at a bakeshop costs $10.20. What is the cost of one muffin?

    (A) $0.85   **(B)** $1.02   **(C)** $0.68   **(D)** $0.75   **(E)** $0.80

PREALGEBRA REVIEW CHAPTER 3 DECIMALS

52. A bowling center charges $3.85 for a single game of bowling. How many games were played if a person pays $11.55?

(A) 5   (B) 3   (C) 6   (D) 2   (E) 4

53. How many square feet of carpeting are needed to carpet a rectangular room that measures 12.2 feet by 16.9 feet?

(A) 20,618   (B) 2601.8   (C) 218.6   (D) 58.2   (E) 206.18

54. An oil company charges $2.41 for each gallon of home heating oil. Last month a customer received a bill for $269.92. How many gallons of oil were delivered?

(A) 121   (B) 1120   (C) 1117   (D) 11.2   (E) 112

55. Owen paid $2.49 for a gallon of gasoline. If it cost him a total of $29.03 to fill his tank, about how many gallons of gasoline did he get?

(A) 11.9   (B) 11.7   (C) 11.6   (D) 11.5   (E) 11.8

56. Fran makes gift baskets for special occasions. It costs $18.54 in supplies to make one basket and she sells each basket for $35.00. What is her profit if she sells 16 baskets?

(A) $560   (B) $296.64   (C) $263.36   (D) $858.24   (E) $16.46

57. An investor bought 500 shares of a stock for $0.86 per share. If she sold the stock when its value increased to $1.98 per share, how much of a profit did the investor make?

(A) $560   (B) $990   (C) $430   (D) $56   (E) $515

58. A fitness center charges $99 to join and a membership fee of $19.99 per month. How much does Sean pay for two years of membership?

(A) $138.98   (B) $677.76   (C) $578.76   (D) $479.76   (E) $338.88

59. Karla makes $6.35 per hour waiting tables at a restaurant. If she worked a 6 hour shift today and made $108.32 in tips, what was the total amount she made today?

(A) $688.02   (B) $146.42   (C) $38.10   (D) $120.57   (E) $147.50

CHAPTER 3   DECIMALS   PREALGEBRA REVIEW

60. Josie bought 0.75 pounds of ham and 1.25 pounds of cheese at the deli. If the ham cost $4.99 per pound and the cheese cost $3.59 per pound, about how much did her deli purchase cost?

   (A) $18.00   (B) $8.90   (C) $8.70   (D) $8.20   (E) $8.60

61. To ride the subway, a commuter can purchase single ride tickets for $1.25 each or a monthly unlimited ride pass for $38. Suppose a commuter purchases the monthly pass and rides the train 42 times this month. How much did he save by buying the monthly pass?

   (A) $52.50   (B) $14.50   (C) $67.00   (D) $8.00   (E) $14.70

62. Maya's company reimburses her $0.32 for every work related mile she drives. Last week she drove a total of 468 miles, 93 of which were not work related. How much should the company reimburse Maya?

   (A) $149.76   (B) $120   (C) $29.76   (D) $155.52   (E) $179.52

63. Justin drove 45.8 miles in the first hour of a trip. If he drove an additional 88.3 miles in the last 1.5 hours of the trip, what was the average number of miles he drove each hour of the trip?

   (A) 81.12   (B) 67.05   (C) 53.64   (D) 89.4   (E) 58.87

64. Through diet and exercise, Tim lost 31.2 pounds in 13 weeks. On average, how many pounds did he lose each week?

   (A) 2.8   (B) 2.5   (C) 2.3   (D) 2.4   (E) 2.9

65. A calling card company charges a $0.35 connection fee for each call and $0.05 for each minute of calling. A customer made 8 calls that totaled 268 minutes. If he had a $25 calling card, how much money does he have left on his card?

   (A) $16.20   (B) $11.25   (C) $7.90   (D) $13.40   (E) $8.80

66. The price of a regular gallon of gas is $2.09 at the self serve island and $2.29 at the full serve island. How much does a customer save on 12 gallons of gas if she uses the self serve island?

   (A) $0.29   (B) $3.48   (C) $2.40   (D) $2.48   (E) $2.70

PREALGEBRA REVIEW

CHAPTER 3 DECIMALS

Section 5: Fractions and Decimals

Converting Decimals to Fractions

Recall from Section 1 that a decimal number can be written in fraction form. This fact provides a way to convert decimals to fractions.

Procedure for Converting a Decimal Number to a Fraction:
Write the decimal number as a decimal fraction. Then simplify the fraction, if possible.

EXAMPLE 1 Convert each decimal number to a fraction.

(a) 0.3   (b) –0.72   (c) 0.125

SOLUTION

(a) $0.3 = \dfrac{3}{10}$

(b) $-0.72 = -\dfrac{72}{100} = -\dfrac{4 \cdot 18}{4 \cdot 25} = -\dfrac{\cancel{4} \cdot 18}{\cancel{4} \cdot 25} = -\dfrac{18}{25}$

(c) $0.125 = \dfrac{125}{1000} = \dfrac{125 \cdot 1}{125 \cdot 8} = \dfrac{\cancel{125} \cdot 1}{\cancel{125} \cdot 8} = \dfrac{1}{8}$

EXAMPLE 2 Convert 3.48 to a mixed number.

SOLUTION

$$3.48 = 3\dfrac{48}{100} = 3\dfrac{12}{25}$$

Converting Fractions to Decimals

Remember that the fraction bar represents division. For instance, $\dfrac{2}{3}$ can be thought of as "2 divided by 3." This fact provides a way to convert a fraction to a decimal number.

Procedure for Converting a Fraction to a Decimal Number:
Write the fraction as a division problem. Then divide.

EXAMPLE 3 Convert the fraction to a decimal number.

(a) $\dfrac{3}{5}$ (b) $\dfrac{9}{16}$ (c) $-\dfrac{7}{20}$

SOLUTION

(a) $\dfrac{3}{5} = 5\overline{)3} \rightarrow 5\overline{)3.0}$ gives 0.6 with remainder 0.

So, $\dfrac{3}{5} = 0.6$

(b) $\dfrac{9}{16} = 16\overline{)9} \rightarrow 16\overline{)9.0000}$ gives 0.5625 with remainder 0.

So, $\dfrac{9}{16} = 0.5625$

(c) Since $-\dfrac{7}{20}$ is negative, divide and make the result negative.

$\dfrac{7}{20} = 20\overline{)7} \rightarrow 20\overline{)7.00}$ gives 0.35 with remainder 0.

So, $-\dfrac{7}{20} = -0.35$

PREALGEBRA REVIEW CHAPTER 3 DECIMALS

EXAMPLE 4   Convert $1\frac{1}{4}$ to a decimal number.

SOLUTION   Write the mixed number as an improper fraction. Then convert the improper fraction to a decimal number.

$$1\frac{1}{4} = \frac{5}{4} \quad \rightarrow \quad \begin{array}{r} 1.25 \\ 4\overline{)5.00} \\ \underline{4} \\ 1\,0 \\ \underline{8} \\ 20 \\ \underline{20} \\ 0 \end{array}$$

EXAMPLE 5   Write $\frac{5}{12}$ as a decimal number.

SOLUTION

$$\frac{5}{12} = \begin{array}{r} 0.41666 \\ 12\overline{)5.00000} \\ \underline{4\,8} \\ 20 \\ \underline{12} \\ 80 \\ \underline{72} \\ 80 \\ \underline{72} \\ 80 \\ \underline{72} \\ 8 \end{array}$$

Notice that no matter how long the division is carried out, the remainder will never be 0. Since 8 keeps reappearing as the remainder, the digit 6 repeats in the quotient. So, $\frac{5}{12} = 0.41666...$ This decimal can also be written as $0.41\overline{6}$, where the bar over 6 indicates the repeating part of the quotient.

If the division results in a remainder of 0, as in Examples 3 and 4, the division is said to *terminate*, and the result is called a *terminating decimal*. If the division does not lead to a remainder of 0, but instead to a repeating pattern of nonzero remainders, as in Example 5, the result is called a *repeating decimal*. Note that every fraction can be written as a decimal number that either terminates or repeats. And every decimal that either terminates or repeats can be written as a fraction.

EXAMPLE 6 Write $-\dfrac{9}{11}$ as a decimal number.

SOLUTION

$$\frac{9}{11} = 11\overline{)9.0000} \begin{array}{c} 0.8181 \\ \end{array}$$

$$\begin{array}{r} 88 \\ \hline 20 \\ 11 \\ \hline 90 \\ 88 \\ \hline 20 \\ 11 \\ \hline 9 \end{array}$$

So, $-\dfrac{9}{11} = -0.8181... = -0.\overline{81}$

Evaluating Expressions Containing Fractions and Decimals

Some expressions may contain both fractions and decimals. These expressions can be evaluated using combinations of the techniques discussed in this chapter and in the preceding chapters.

EXAMPLE 7 Evaluate: $0.36 \times \dfrac{1}{2}$

SOLUTION

$$0.36 \times \frac{1}{2} = \frac{0.36}{1} \times \frac{1}{2} = \frac{0.36 \times 1}{1 \times 2} = \frac{0.36}{2} = 0.18$$

PREALGEBRA REVIEW   CHAPTER 3   DECIMALS

EXAMPLE 8   Calculate.

$$\text{(a) } \frac{1}{4}(1.2+0.4) \quad \text{(b) } \frac{3}{5}\times 2.5 - 0.24 \times \frac{5}{12}$$

SOLUTION

(a) $\frac{1}{4}(1.2+0.4) = \frac{1}{4}(1.6)$

$= \frac{1}{4}\left(\frac{1.6}{1}\right)$

$= \frac{1.6}{4}$

$= 0.4$

(b) $\frac{3}{5}\times 2.5 - 0.24 \times \frac{5}{12} = \frac{3}{5}\times\frac{2.5}{1} - \frac{0.24}{1}\times\frac{5}{12}$

$= \frac{7.5}{5} - \frac{1.20}{12}$

$= 1.5 - 0.1$

$= 1.4$

EXAMPLE 9   Evaluate: $5.69 + \frac{1}{6} \div \frac{2}{3} \cdot 8.64$

SOLUTION

$5.69 + \frac{1}{6} \div \frac{2}{3} \cdot 8.64 = 5.69 + \frac{1}{6} \cdot \frac{3}{2} \cdot 8.64$

$= 5.69 + \frac{1 \cdot \cancel{3}^{1}}{2 \cdot \cancel{3} \cdot 2} \cdot 8.64$

$= 5.69 + \frac{1}{4} \cdot \frac{8.64}{1}$

$= 5.69 + \frac{8.64}{4}$

$= 5.69 + 2.16$

$= 7.85$

CHAPTER 3   DECIMALS   PREALGEBRA REVIEW

Section 5: Exercises

Convert the decimal number to a fraction or mixed number.

1. 0.1
2. 0.5
3. 0.75
4. 0.48

5. −0.6
6. −0.7
7. −0.64
8. −0.55

9. 1.2
10. 4.24
11. 0.256
12. −0.475

13. −2.016
14. 7.875
15. 0.3125
16. 0.9375

Convert the fraction or mixed number to a decimal number.

17. $\dfrac{2}{5}$
18. $\dfrac{5}{8}$
19. $-\dfrac{1}{4}$
20. $-\dfrac{1}{20}$

21. $3\dfrac{1}{2}$
22. $2\dfrac{7}{8}$
23. $\dfrac{9}{25}$
24. $\dfrac{29}{32}$

25. $-1\dfrac{11}{16}$
26. $\dfrac{21}{40}$
27. $\dfrac{88}{125}$
28. $-\dfrac{149}{200}$

29. $\dfrac{2}{3}$
30. $\dfrac{2}{9}$
31. $-\dfrac{1}{6}$
32. $-\dfrac{10}{11}$

33. $2\dfrac{1}{3}$
34. $1\dfrac{1}{12}$
35. $\dfrac{13}{15}$
36. $-\dfrac{7}{18}$

37. $-\dfrac{9}{22}$
38. $\dfrac{25}{36}$
39. $\dfrac{32}{45}$
40. $\dfrac{99}{111}$

41. $-3\dfrac{8}{11}$
42. $-5\dfrac{4}{15}$
43. $\dfrac{2}{7}$
44. $\dfrac{9}{13}$

Calculate.

45. $\dfrac{2}{3} \times 2.16$
46. $5.6\left(\dfrac{5}{7}\right)$
47. $\dfrac{4}{5}(8.93 - 4.08)$

48. $\dfrac{1}{6}(2.1+3.3)$ **49.** $0.8\times\dfrac{7}{10}+0.04$ **50.** $9.1-\dfrac{8}{9}\cdot 1.08$

51. $2\dfrac{1}{5}-1.99$ **52.** $3\dfrac{1}{8}+0.025$ **53.** $0.3\left(\dfrac{5}{6}\right)+10.15\left(\dfrac{3}{5}\right)$

54. $1\dfrac{1}{9}\times 0.045-0.36\times\dfrac{1}{12}$ **55.** $\dfrac{9}{16}-\dfrac{3}{4}(3.7-1.3)\times\dfrac{1}{2}$ **56.** $1.6\times\dfrac{7}{8}+\dfrac{1}{3}(6.6+2.4)$

Choose the correct answer.

57. During a sale, a sporting goods store advertises $\dfrac{1}{3}$ off all baseball equipment. What is the sale price of a baseball glove with an original price of $54.99?

   (A) $37.66   **(B)** $18.33   **(C)** $35.99   **(D)** $36.66   **(E)** $27.33

58. A pound of coffee costs $4.90. How much does it cost for $2\dfrac{1}{2}$ pounds of coffee?

   (A) $11.25   **(B)** $9.80   **(C)** $12.50   **(D)** $12.25   **(E)** $11.75

59. A taxi company charges $2.00 for the first $\dfrac{1}{2}$ mile of a trip and $0.35 for each additional $\dfrac{1}{4}$ mile. If a fare was $16.35, how many miles did the taxi drive?

   (A) 10   **(B)** $10\dfrac{3}{4}$ **(C)** $10\dfrac{1}{4}$ **(D)** 41   **(E)** $10\dfrac{1}{2}$

60. A parking garage charges $2.50 for the first hour and $.75 for each additional $\dfrac{1}{2}$ hour up to 8 hours. After 8 hours the garage charges a flat fee of $15.00. How much will a person pay to park their car in the garage for $5\dfrac{1}{2}$ hours?

   (A) $9.25   **(B)** $5.50   **(C)** $10.75   **(D)** $15.00   **(E)** $8.25

CHAPTER 3 DECIMALS   PREALGEBRA REVIEW

61. A factory worker earns $1\frac{1}{2}$ times his hourly pay rate of $12.70 for any hours worked over 40 hours. In addition, he earns twice his pay rate for working on weekends. If he worked 46.8 hours during the week (Monday through Friday) and 4 hours on Saturday, what were earnings for the week?

   (A) $645.16   **(B)** $739.14   **(C)** $899.54   **(D)** $688.34   **(E)** $967.74

62. During a clearance sale, a sweater was marked $\frac{3}{5}$ off. An additional $5.80 was taken off the sale price at the register. If the final sale price was $17.40, what was the original price of the sweater?

   (A) $58.00   **(B)** $38.67   **(C)** $43.50   **(D)** $60.00   **(E)** $23.20

PREALGEBRA REVIEW

CHAPTER 3   DECIMALS

Section 6: Scientific Notation

Properties of Exponents

In preceding chapters, the Order of Operations was used to simplify numerical expressions containing multiple operations. For instance, $2^3 \cdot 5^2$ can be simplified as follows:

$$2^3 \cdot 5^2 = 8 \cdot 25 \quad \textit{Evaluate expressions containing exponents.}$$
$$= 200 \quad \textit{Multiply.}$$

EXAMPLE 1   Evaluate.

(a) $4^3 \cdot 3^2$ (b) $8^2 \div 2^4$ (c) $\left(2^4\right)^2$

SOLUTION

(a) $4^3 \cdot 3^2 = 64 \cdot 9 = 576$
(b) $8^2 \div 2^4 = 64 \div 16 = 4$
(c) $\left(2^4\right)^2 = (16)^2 = 256$

In some situations, a product of two numbers can be written directly in exponent form. For instance, consider the product $6^2 \cdot 6^3$. Notice that both factors have the *same* base, 6. To write this product in exponent form, first write it in expanded form and then rewrite it in exponent form.

$$6^2 \cdot 6^3 = \underbrace{6 \cdot 6}_{\text{2 factors of 6}} \cdot \underbrace{6 \cdot 6 \cdot 6}_{\text{3 factors of 6}} = 6^5$$

Notice that the exponent of the product, 5, is the sum of the exponents, 2 and 3, of the original factors. That is, $6^2 \cdot 6^3 = 6^{2+3} = 6^5$. When the base is the same, the product can be written directly in exponent form by simply adding the exponents. This is generalized in the *Product Rule of Exponents*.

Product Rule of Exponents: For any nonzero number a, and integers m and n, $a^m \cdot a^n = a^{m+n}$.

According to this rule, when multiplying expressions with the same base, add the exponents and keep the base the same.

CHAPTER 3 DECIMALS

Remember by the order of operations, exponents are evaluated before multiplying. So *do not multiply the bases* before applying the Product Rule of Exponents.

EXAMPLE 2 Simplify.

(a) $2^5 \cdot 2^7$ (b) $(-5)^6 \cdot (-5)^1$ (c) $10^3 \cdot 10^4$

SOLUTION

(a) $2^5 \cdot 2^7 = 2^{5+7} = 2^{12}$
(b) $(-5)^6 \cdot (-5)^1 = (-5)^{6+1} = (-5)^7$
(c) $10^3 \cdot 10^4 = 10^{3+4} = 10^7$

As with the product of two numbers with the same base, the quotient of two numbers with the same base can also be written directly in exponent form. For example, to write the quotient $\dfrac{6^5}{6^3}$ in exponent form, first write the numerator and denominator in expanded form and simplify.

$$\frac{6^5}{6^3} = \frac{6 \cdot 6 \cdot 6 \cdot 6 \cdot 6}{6 \cdot 6 \cdot 6} = \frac{\cancel{6} \cdot \cancel{6} \cdot \cancel{6} \cdot 6 \cdot 6}{\cancel{6} \cdot \cancel{6} \cdot \cancel{6}} = \frac{6 \cdot 6}{1} = \frac{6^2}{1} = 6^2$$

Notice that the exponent of the quotient, 2, is the difference of the exponents, 5 and 3, of the dividend and divisor. That is, $\dfrac{6^5}{6^3} = 6^{5-3} = 6^2$. When the base is the same, the quotient can be written directly in exponent form by simply subtracting the exponents. This is generalized in the *Quotient Rule of Exponents*.

Quotient Rule of Exponents: For any nonzero number a, and integers m and n, $\dfrac{a^m}{a^n} = a^{m-n}$.

According to this rule, when dividing expressions with the same base, subtract the exponents and keep the base the same. Note that the order of subtraction is important. The exponent in the denominator (or divisor) is always subtracted from the exponent in the numerator (or dividend).

Remember by the order of operations, exponents are evaluated before dividing. So *do not divide the bases* before applying the Quotient Rule of Exponents.

PREALGEBRA REVIEW   CHAPTER 3 DECIMALS

EXAMPLE 3   Simplify.

$$\text{(a) } \frac{3^6}{3^1} \qquad \text{(b) } (-4)^7 \div (-4)^4 \qquad \text{(c) } \frac{10^{11}}{10^8}$$

SOLUTION

(a) $\dfrac{3^6}{3^1} = 3^{6-1} = 3^5$

(b) $(-4)^7 \div (-4)^4 = (-4)^{7-4} = (-4)^3$

(c) $\dfrac{10^{11}}{10^8} = 10^{11-8} = 10^3$

Note that if you forget the Product and Quotient Rules, you can still simplify by first expanding the expressions, simplifying if necessary, and then rewriting in exponent form.

Negative Exponents

So far in this book, operations with positive exponents have been considered. The concept of an exponent can be expanded to include negative exponents. A negative exponent is used to denote a power that is in the denominator of a fraction. For instance, $\dfrac{1}{4^3}$ can be written using a negative exponent as 4^{-3}.

Negative Exponent: For any nonzero number a and positive integer n, $\dfrac{1}{a^n} = a^{-n}$.

EXAMPLE 4   Simplify.

$$\text{(a) } 2^{-4} \qquad \text{(b) } (-3)^{-3}$$

SOLUTION

(a) $2^{-4} = \dfrac{1}{2^4} = \dfrac{1}{16}$

(b) $(-3)^{-3} = \dfrac{1}{(-3)^3} = \dfrac{1}{-27} = -\dfrac{1}{27}$

CHAPTER 3   DECIMALS

Scientific Notation

Recall from Section 4 that a number multiplied by a power of 10, moves the decimal point in the original number to the *right* the same number of places as the number of zeros in the power of 10. For instance, $5.6 \times 10,000,000 = 56,000,000$.

Similarly, a number divided by a power of 10, moves the decimal point in the original number to the *left* the same number of places as the number of zeros in the power of 10. For instance, $\frac{5.6}{100,000} = 0.000056$.

Scientific notation is a way to write very large or very small numbers in a condensed form. For instance, the number 56,000,000 can be written in scientific notation as 5.6×10^7. (Note that $5.6 \times 10^7 = 5.6 \times 10,000,000 = 56,000,000$.) A number is written in scientific notation if it is written as the product of a number whose absolute value is greater than or equal to 1 and less than 10 and a power of 10.

> **Scientific Notation:** A number written in scientific notation has the form
> $$a \times 10^n,$$
> where $1 \leq |a| < 10$ and n is an integer.

EXAMPLE 5   Determine if the number is written in scientific notation.

(a) 35.7×10^4 (b) 2.3×10^6 (c) -1.722×10^8

SOLUTION

(a) The number is not written in scientific notation because $|35.7| = 35.7$ is not between 1 and 10.

(b) The number is written in scientific notation.

(c) The number is written in scientific notation.

To convert a number to scientific notation, write it in the form $a \times 10^n$. For example, to convert 230,000,000 to scientific notation, rewrite it as the product of a number between 1 and 10 and a power of 10. To do this, move the decimal point *to the left* 8 places so that it is placed between the 2 and the 3, giving 2.30000000. Then drop the trailing zeros and multiply this number by 10^8, a power of 10 with exponent that is equal to the number of places the decimal point was moved. So
$$230,000,000 = 2.3 \times 10^8$$

PREALGEBRA REVIEW   CHAPTER 3   DECIMALS

EXAMPLE 6   Convert 1,245,000,000,000 to scientific notation.

SOLUTION $1{,}245{,}000{,}000{,}000 = 1.245 \times 10^{12}$

Now consider a very small number, such as 0.0000000056. This number can be written in scientific notation as 5.6×10^{-9}. Here a negative exponent is used as the power of 10. (Note that $5.6 \times 10^{-9} = 5.6 \times \dfrac{1}{10^9} = \dfrac{5.6}{10^9} = \dfrac{5.6}{1{,}000{,}000{,}000} = 0.0000000056$.) In this case, notice that the decimal point is moved 9 places *to the right*.

EXAMPLE 7   Convert 0.000315 to scientific notation.

SOLUTION $0.000315 = 3.15 \times 10^{-4}$

Procedure for Converting a Number in Standard Notation to Scientific Notation:
Move the decimal point to the right of the first nonzero digit. Then multiply the number by 10^n, where n is the number of places the decimal point was moved. If the decimal point was moved to the left, n is positive. If n was moved to the right, n is negative.

EXAMPLE 8   Convert to scientific notation.

      (a) 8300   (b) 0.0000000000012   (c) –72,000,000   (d) –0.009

SOLUTION
(a) $8300 = 8.3 \times 10^3$
(b) $0.0000000000012 = 1.2 \times 10^{-12}$
(c) $-72{,}000{,}000 = -7.2 \times 10^7$
(d) $-0.009 = -9 \times 10^{-3}$

To convert a number from scientific notation to standard notation, the process is reversed.

Procedure for Converting a Number in Scientific Notation to Standard Notation:
- *Positive power of 10* – Move the decimal point to the right the same number of places as the exponent, inserting zeros as necessary.
- *Negative power of 10* – Move the decimal point to the left the same number of places as the exponent, inserting zeros as necessary.

CHAPTER 3 DECIMALS PREALGEBRA REVIEW

EXAMPLE 9 Write the number in standard notation.

(a) 6.8×10^6 (b) 5.75×10^{-8}

SOLUTION

(a) $6.8 \times 10^6 = 6,800,000$ *Move the decimal point 6 places to the right.*

(b) $5.75 \times 10^{-8} = 0.0000000575$ *Move the decimal point 8 places to the left.*

Multiplication and Division with Scientific Notation

When multiplying or dividing very large or very small numbers, scientific notation along with the product and quotient rules of exponents can be used to make computations simpler. For instance, consider the product of 200,000 and 42,000. Converting each number to scientific notation and multiplying gives

$$200,000 \times 42,000 = (2 \times 10^5) \times (4.2 \times 10^4)$$ *Convert factors to scientific notation.*
$$= (2 \times 4.2) \times (10^5 \times 10^4)$$ *Use the commutative and associative properties.*
$$= 8.4 \times 10^{5+4}$$ *Apply the product rule of exponents.*
$$= 8.4 \times 10^9$$

EXAMPLE 10 Multiply. Express the product in scientific notation

(a) $(1.8 \times 10^2) \cdot (5.1 \times 10^4)$ (b) $(3.3 \times 10^{-6})(2.1 \times 10^3)$

SOLUTION

(a) $(1.8 \times 10^2) \cdot (5.1 \times 10^4) = (1.8 \times 5.1) \times (10^2 \times 10^4) = 9.18 \times 10^{2+4} = 9.18 \times 10^6$

(b) $(3.3 \times 10^{-6})(2.1 \times 10^3) = (3.3 \times 2.1) \times (10^{-6} \times 10^3) = 6.93 \times 10^{-6+3} = 6.93 \times 10^{-3}$

EXAMPLE 11 Divide. Express the quotient in scientific notation.

(a) $\dfrac{9.6 \times 10^9}{2.4 \times 10^4}$ (b) $\dfrac{2.66 \times 10^3}{1.4 \times 10^{-5}}$

SOLUTION

(a) $\dfrac{9.6 \times 10^9}{2.4 \times 10^4} = \dfrac{9.6}{2.4} \times \dfrac{10^9}{10^4} = 4 \times 10^{9-4} = 4 \times 10^5$

(b) $\dfrac{2.66 \times 10^3}{1.4 \times 10^{-5}} = \dfrac{2.66}{1.4} \times \dfrac{10^3}{10^{-5}} = 1.9 \times 10^{3-(-5)} = 1.9 \times 10^8$

PREALGEBRA REVIEW   CHAPTER 3   DECIMALS

EXAMPLE 12   Compute and write the answer in scientific notation.

   (a) $48,000,000 \times 20,000$ **(b)** $\dfrac{850,000}{170,000,000}$ **(c)** $12,400,000 \cdot 0.0007$

SOLUTION

   (a) $48,000,000 \times 20,000 = (4.8 \times 10^7) \times (2 \times 10^4) = 9.6 \times 10^{11}$

   (b) $\dfrac{850,000}{170,000,000} = \dfrac{8.5 \times 10^5}{1.7 \times 10^8} = 5 \times 10^{-3}$

   (c) $12,400,000 \cdot 0.0007 = (1.24 \times 10^7) \cdot (7 \times 10^{-4}) = 8.68 \times 10^3$

EXAMPLE 13   Calculate $(4.5 \times 10^{10}) \cdot (2.2 \times 10^{-5})$. Then write the answer in standard notation.

SOLUTION $(4.5 \times 10^{10}) \cdot (2.2 \times 10^{-5}) = 9.9 \times 10^5 = 990,000$

EXAMPLE 14   Calculate and write the answer in scientific notation.

   (a) $(1.4 \times 10^3)^2$ **(b)** $\dfrac{(2.5 \times 10^{-2}) \cdot (3.6 \times 10^{12})}{3 \times 10^4}$

SOLUTION

   (a) $(1.4 \times 10^3)^2 = (1.4 \times 10^3)(1.4 \times 10^3) = 1.96 \times 10^6$

   (b) $\dfrac{(2.5 \times 10^{-2}) \cdot (3.6 \times 10^{12})}{3 \times 10^4} = \dfrac{9 \times 10^{10}}{3 \times 10^4} = 3 \times 10^6$

Section 6: Exercises

Evaluate.

1. $3^3 \cdot 4$
2. $5^2 \cdot 6^2$
3. $\dfrac{9^2}{3^4}$

4. $12^2 \div 2^4$
5. $\left(3^3\right)^2$
6. $\dfrac{\left(4^2\right)^2}{16}$

Simplify.

7. $9^3 \cdot 9^8$
8. $(-3)^{12} \cdot (-3)^5$
9. $7^7 \div 7^2$

10. $\dfrac{(-6)^{13}}{(-6)^9}$
11. $3^6 \cdot 3^1 \cdot 3^{10}$
12. $\dfrac{4^4 \cdot 4^8}{4^{11}}$

Write using a positive exponent, then simplify.

13. 3^{-4}
14. 6^{-2}
15. $(-2)^{-5}$
16. $(-7)^{-2}$

Convert the number to scientific notation.

17. 90,000

18. 670,000,000

19. 0.0000000006

20. 0.00000551

21. −8,893,000,000,000

22. −420,000

23. −0.00037

24. −0.00000002269

Convert the number to standard form.

25. 4×10^7

26. 7.5×10^{10}

27. 6.8×10^{-4}

28. 3.38×10^{-9}

PREALGEBRA REVIEW   CHAPTER 3 DECIMALS

29. -1.99×10^6 **30.** -9.2×10^{-2}

Calculate. Write the answer in scientific notation.

31. $(3 \times 10^4)(2 \times 10^6)$ **32.** $(1.4 \times 10^7)(4.9 \times 10^7)$

33. $(2.3 \times 10^{-5})(4 \times 10^{13})$ **34.** $(1.22 \times 10^6)(8 \times 10^{-9})$

35. $\dfrac{8 \times 10^{10}}{2 \times 10^7}$ **36.** $\dfrac{7.2 \times 10^{12}}{6 \times 10^4}$

37. $\dfrac{2.25 \times 10^{-6}}{1.5 \times 10^5}$ **38.** $\dfrac{8.84 \times 10^9}{2.6 \times 10^{-9}}$

39. $40,000 \cdot 700$ **40.** $62,000,000 \times 1,000,000,000$

41. $\dfrac{450,000,000}{150,000}$ **42.** $\dfrac{36,000}{1,200,000,000}$

43. $(23,000,000)(0.00005)$ **44.** $0.0048 \cdot 0.0000002$

45. $\dfrac{0.000072}{16,000,000}$ **46.** $\dfrac{5,600,000}{0.00000014}$

47. $\left(2.5 \times 10^5\right)^2$ **48.** $\left(1.1 \times 10^8\right)^2$

49. $\dfrac{8.8 \times 10^{15}}{(2 \times 10^7)(4 \times 10^3)}$ **50.** $\dfrac{(5.6 \times 10^6)(1.6 \times 10^8)}{6.4 \times 10^9}$

51. $\dfrac{(2.8 \times 10^{-7})(3 \times 10^8)}{(1.2 \times 10^2)(7 \times 10^{-10})}$ **52.** $\dfrac{\left(2.4 \times 10^4\right)^2}{(4 \times 10^{-1})(1.2 \times 10^{-9})}$

Compute. Write the answer in standard notation.

53. $(5000)(40,000)$ **54.** $85,000,000 \times 0.00004$

CHAPTER 3   DECIMALS   PREALGEBRA REVIEW

55. $(2.3 \times 10^{-8})(4 \times 10^{5})$

56. $(1.25 \times 10^{9})(6.6 \times 10^{-7})$

57. $\dfrac{7.6 \times 10^{6}}{4 \times 10^{2}}$

58. $\dfrac{1.69 \times 10^{-10}}{1.3 \times 10^{-5}}$

Choose the correct answer.

59. Which of the following numbers is written in scientific notation?

    (A) 12.2×10^{4} **(B)** 0.122×10^{3} **(C)** 1.22×10^{4}
    (D) 122×10^{2} **(E)** 12.2

60. Which of the following numbers is not written in scientific notation?

    (A) -2.44×10^{-2} **(B)** 1.3×10^{8} **(C)** -9.99×10^{3}
    (D) 0.59×10^{-3} **(E)** 4.3×10^{1}

61. The product of $200{,}000$ and (8.2×10^{9}) in scientific notation is

    (A) 16.4×10^{14} **(B)** 4.1×10^{5} **(C)** 1.64×10^{14}
    (D) 1.64×10^{13} **(E)** 1.64×10^{15}

62. Divide $64{,}000$ by (1.6×10^{6}) and write the answer in standard form.

    (A) $102{,}400{,}000{,}000$ **(B)** 0.04 **(C)** 25 **(D)** 0.004 **(E)** 4×10^{-2}

63. Multiply: $(-2.1 \times 10^{7})(3 \times 10^{-5})$

    (A) -6.3×10^{2} **(B)** -6.3×10^{12} **(C)** 6.3×10^{2}
    (D) -6.3×10^{-2} **(E)** 6.3×10^{-2}

64. Divide: $\dfrac{3.6 \times 10^{10}}{7.2 \times 10^{14}}$

    (A) 0.5×10^{-4} **(B)** 5×10^{24} **(C)** 5×10^{-4} **(D)** 5×10^{-5} **(E)** 2×10^{-4}

65. The mass of Earth is about 6×10^{24} kilograms. The mass of Earth's moon is about 0.0123 times the mass of Earth. What is the mass of Earth's moon?

    (A) 7.38×10^{22} kg   **(B)** 7.38×10^{24} kg   **(C)** 7.38×10^{-22} kg
    (D) 4.88×10^{26} kg   **(E)** 7.38×10^{26} kg

66. A carbon atom weighs 2×10^{-23} grams. How much do 10,000,000 carbon atoms weigh?

    (A) 2×10^{16} g   **(B)** 2×10^{-30} g   **(C)** 2×10^{-16} g
    (D) 2×10^{-17} g   **(E)** 0.2×10^{-16} g

67. The distance from the Sun to Neptune is about 4,500,000,000 kilometers. If one astronomical unit is about 150,000,000 kilometers, how many astronomical units is Neptune from the Sun?

    (A) 3   **(B)** 300   **(C)** 0.3   **(D)** 30   **(E)** 0.03

68. The surface area of Mercury is about 7.5×10^{7} square kilometers. The surface area of the Sun is about 6.09×10^{12} kilograms. How many times the surface area of Mercury is the surface area of the Sun?

    (A) 0.812   **(B)** 81,200   **(C)** 812,000   **(D)** 8120   **(E)** 81.2

CHAPTER 3   DECIMALS

Section 7: Square Roots

Perfect Squares

When a number is raised to the power 2, the number is said to be **squared** and the result is called a **perfect square**. For instance, 3^2, read as "three squared", is equal to 9. The number 9 is a perfect square.

EXAMPLE 1   Determine whether the number is a perfect square.

(a) 8   (b) 49   (c) $\dfrac{9}{16}$ (d) –4

SOLUTION

(a) 8 is not a perfect square since there is no number (integer, fraction, or decimal) that when squared is 8.

(b) 49 is a perfect square since $7^2 = 7 \cdot 7 = 49$.

(c) $\dfrac{9}{16}$ is a perfect square since $\left(\dfrac{3}{4}\right)^2 = \dfrac{3}{4} \cdot \dfrac{3}{4} = \dfrac{9}{16}$. (Note that both the numerator and denominator contain perfect squares.)

(d) –4 is not a perfect square since there is no number that when squared yields a negative number.

Some perfect squares are listed below.

$1^2 = 1$ $6^2 = 36$
$2^2 = 4$ $7^2 = 49$
$3^2 = 9$ $8^2 = 64$
$4^2 = 16$ $9^2 = 81$
$5^2 = 25$ $10^2 = 100$

Note that these squares are the same when the base is negative as well. For example, $(-1)^2 = 1$, $(-2)^2 = 4$, $(-3)^2 = 9$, $(-4)^2 = 16$, and so on.

PREALGEBRA REVIEW CHAPTER 3 DECIMALS

Square Roots of Perfect Squares

If a number is the product of a factor times itself, the factor is called a **square root** of the number. For example, since $3^2 = \underbrace{3 \cdot 3}_{\text{The factor 3 times itself}} = 9$, 3 is a square root of 9.

Every positive square number has two square roots, one positive and one negative. For example, the square roots of 9 are 3 and –3, since $3^2 = 9$ and $(-3)^2 = 9$. To indicate which root is needed, the **radical symbol**, $\sqrt{}$, read as "the square root of" or "radical," is used to denote the positive square root of the number inside the radical, which is called the **radicand.** For instance,

$$\underset{\substack{\uparrow \\ \text{9 is the} \\ \text{radicand}}}{\sqrt{9}} = 3 \qquad \text{"The square root of 9 is 3."}$$

When the negative square root of a number is needed, the radical symbol is preceded by a negative sign, $-\sqrt{}$. For instance,

$$-\sqrt{9} = -(3) = -3 \qquad \text{"The negative square root of 9 is –3."}$$

Note that 0 is the only number with one square root: $\sqrt{0} = 0$ because $0^2 = 0$

EXAMPLE 2   Find the square root of the number.

   (a) $\sqrt{25}$ (b) $-\sqrt{64}$ (c) $\sqrt{\dfrac{1}{16}}$ (d) $\sqrt{0.04}$ (e) $\sqrt{3600}$

SOLUTION

(a) $\sqrt{25} = 5$, since $5^2 = 25$.

(b) $-\sqrt{64} = -(8) = -8$, since –8 is the negative square root of 64.

(c) $\sqrt{\dfrac{1}{16}} = \dfrac{1}{4}$, since $\left(\dfrac{1}{4}\right)^2 = \dfrac{1}{16}$. (Note that both the numerator and the denominator of the radicand are perfect squares.)

(d) $\sqrt{0.04} = 0.2$, since $(0.2)^2 = 0.04$.

(e) $\sqrt{3600} = 60$, since $60^2 = 3600$

Approximating Square Roots

Square roots with radicands that are not perfect squares cannot be written as whole numbers, fractions, or terminating or repeating decimals. For instance, $\sqrt{2}$, $\sqrt{5}$, and $\sqrt{10}$ are square roots that do not have exact decimal representations. However, the value of these square roots can be approximated. To see this, consider $\sqrt{2}$.

Notice that the radicand, 2, lies between the two perfect squares 1 and 4 on the number line. This means that $\sqrt{2}$ will fall between $\sqrt{1}$ and $\sqrt{4}$, or 1 and 2 respectively. That is, $1 < \sqrt{2} < 2$.

$\sqrt{2} \approx 1.4$, since $(1.4)^2 = 1.96$

$\sqrt{2} \approx 1.41$, since $(1.41)^2 = 1.9881$

$\sqrt{2} \approx 1.414$, since $(1.414)^2 = 1.99396$

$\sqrt{2} \approx 1.4142$, since $(1.4142)^2 = 1.99996164$

The symbol $\approx$ means "is approximately equal to."

Notice that each decimal approximation is closer to $\sqrt{2}$, however, none is exactly $\sqrt{2}$.

In general, when the radicand of a square root is not a perfect square, the square root is a non-terminating, non-repeating decimal. Such numbers are called **irrational numbers.**

EXAMPLE 3 Find two consecutive whole numbers for which the exact value of the square root lies between.

(a) $\sqrt{11}$ (b) $\sqrt{45}$

SOLUTION

(a) Since 11 lies between the two perfect squares 9 and 16 on the number line and since $\sqrt{9} = 3$ and $\sqrt{16} = 4$, $\sqrt{11}$ lies between 3 and 4. This can be written as $3 < \sqrt{11} < 4$, which is read as "three is less than the square root of eleven is less than four."

(b) Since 45 lies between the two perfect squares 36 and 49 on the number line and since $\sqrt{36} = 6$ and $\sqrt{49} = 7$, $\sqrt{45}$ lies between 6 and 7. This can be written as $6 < \sqrt{45} < 7$.

Note that the integers together with fractions and terminating and repeating decimals make up a set of numbers called **rational numbers.** The *rational numbers* together with the *irrational numbers* make up a set of numbers called **real numbers.**

PREALGEBRA REVIEW

CHAPTER 3   DECIMALS

Simplifying Expressions with Square Roots

In the order of operations agreement all operations inside grouping symbols such as parentheses, brackets, and absolute values should be done first. The radical symbol can also be considered a grouping symbol. When an expression contains a square root, perform any operations within the square root, and then take the square root.

EXAMPLE 4   Simplify: $\sqrt{15-11}$

SOLUTION $\sqrt{15-11} = \sqrt{4} = 2$

EXAMPLE 5   Evaluate: $\sqrt{16}+\sqrt{81}$

SOLUTION $\sqrt{16}+\sqrt{81} = 4+9 = 13$

EXAMPLE 6   Simplify.

(a) $3\sqrt{49}$ (b) $\sqrt{36}\left(\sqrt{9}\right)$

SOLUTION

(a) $3\sqrt{49} = 3 \cdot \sqrt{49} = 3 \cdot 7 = 21$
(b) $\sqrt{36}\left(\sqrt{9}\right) = 6(3) = 18$

When multiplying two radicals with radicands that are perfect squares, such as $\sqrt{36}$ and $\sqrt{9}$, the square roots can be evaluated before multiplying, as in Example 6(b). However, when multiplying two radicals with radicands that are not perfect squares, such as $\sqrt{8} \cdot \sqrt{32}$, the product of the square roots can be calculated first, and then the result simplified, if possible. To multiply two radicals, use the *product rule for radicals*.

Product Rule for Radicals: If a and b are nonnegative numbers, then $\sqrt{a} \cdot \sqrt{b} = \sqrt{a \cdot b}$

CHAPTER 3   DECIMALS

EXAMPLE 7   Multiply. Simplify, if possible.

(a) $\sqrt{8} \cdot \sqrt{18}$ (b) $\sqrt{27} \cdot \sqrt{3}$ (c) $\sqrt{5} \cdot \sqrt{6}$

SOLUTION

(a) $\sqrt{8} \cdot \sqrt{18} = \sqrt{8 \cdot 18} = \sqrt{144} = 12$
(b) $\sqrt{27} \cdot \sqrt{3} = \sqrt{27 \cdot 3} = \sqrt{81} = 9$
(c) $\sqrt{5} \cdot \sqrt{6} = \sqrt{5 \cdot 6} = \sqrt{30}$

EXAMPLE 8   Simplify: $\left(\sqrt{7}\right)^2$

SOLUTION $\left(\sqrt{7}\right)^2 = \sqrt{7} \cdot \sqrt{7} = \sqrt{7 \cdot 7} = \sqrt{49} = 7$

In Example 8, notice that the result of squaring $\sqrt{7}$ is simply the radicand, 7. In general, for any nonnegative number a, $\left(\sqrt{a}\right)^2 = a$.

EXAMPLE 9   Simplify.

(a) $\left(\sqrt{15}\right)^2$ (b) $3^2 + \left(\sqrt{10}\right)^2$

SOLUTION

(a) $\left(\sqrt{15}\right)^2 = 15$
(b) $3^2 + \left(\sqrt{10}\right)^2 = 9 + 10 = 19$

EXAMPLE 10   Evaluate.

(a) $\dfrac{\sqrt{81}}{\sqrt{9}}$ (b) $\sqrt{\dfrac{32}{2}}$

SOLUTION

(a) $\dfrac{\sqrt{81}}{\sqrt{9}} = \dfrac{9}{3} = 3$
(b) $\sqrt{\dfrac{32}{2}} = \sqrt{16} = 4$

When dividing two radicals with radicands that are perfect squares, such as $\sqrt{81}$ and $\sqrt{9}$, the square roots can be evaluated before dividing or simplifying, as in Example 10(a). However, when dividing two radicals with radicands that are not perfect squares, such as $\dfrac{\sqrt{32}}{\sqrt{8}}$, the quotient of the square roots can be calculated first, and then the result simplified, if possible. To divide two radicals, use the *quotient rule for radicals*

Quotient Rule for Radicals: If a and b are nonnegative numbers, $\dfrac{\sqrt{a}}{\sqrt{b}} = \sqrt{\dfrac{a}{b}}$.

The quotient rule for radicals is especially useful when dividing two square roots in which one or both radicands are not perfect squares.

EXAMPLE 11 Evaluate: $\dfrac{\sqrt{54}}{\sqrt{6}}$

SOLUTION Both the numerator and the denominator contain square roots with radicands that are not perfect squares. To evaluate, rewrite the expression using the quotient rule for radicals.

$$\dfrac{\sqrt{54}}{\sqrt{6}} = \sqrt{\dfrac{54}{6}} \qquad \textit{Rewrite the quotient.}$$
$$= \sqrt{9} \qquad \textit{Divide 54 by 6.}$$
$$= 3 \qquad \textit{Simplify.}$$

EXAMPLE 12 Simplify: $\dfrac{\sqrt{21}}{\sqrt{7}}$

SOLUTION $\dfrac{\sqrt{21}}{\sqrt{7}} = \sqrt{\dfrac{21}{7}} = \sqrt{3}$

Section 7: Exercises

Evaluate.

1. $\sqrt{4}$
2. $\sqrt{36}$
3. $\sqrt{121}$

4. $\sqrt{225}$
5. $-\sqrt{16}$
6. $-\sqrt{100}$

7. $\sqrt{\dfrac{25}{36}}$
8. $\sqrt{\dfrac{49}{81}}$
9. $-\sqrt{\dfrac{1}{25}}$

10. $-\sqrt{\dfrac{4}{9}}$
11. $-\sqrt{1.69}$
12. $\sqrt{0.64}$

13. $\sqrt{0.09}$
14. $-\sqrt{2.56}$
15. $\sqrt{400}$

16. $-\sqrt{625}$
17. $-\sqrt{8100}$
18. $\sqrt{1600}$

Find two consecutive whole numbers for which the exact value of the square root lies.

19. $\sqrt{19}$
20. $\sqrt{50}$
21. $\sqrt{110}$
22. $\sqrt{99}$

Simplify.

23. $\sqrt{28+21}$
24. $\sqrt{96-32}$
25. $\sqrt{81}-\sqrt{4}$

26. $\sqrt{100}+\sqrt{16}$
27. $-\sqrt{81}+\sqrt{49}$
28. $-\sqrt{144}+\left(-\sqrt{16}\right)$

29. $5\sqrt{9}$
30. $6\sqrt{36}$
31. $-2\sqrt{25}$

32. $-4\sqrt{121}$
33. $\sqrt{25}\cdot\sqrt{49}$
34. $\sqrt{1}\cdot\sqrt{9}$

35. $\left(-\sqrt{16}\right)\left(\sqrt{64}\right)$
36. $\left(-\sqrt{100}\right)\left(-\sqrt{169}\right)$
37. $\sqrt{28}\cdot\sqrt{7}$

38. $\sqrt{5}\cdot\sqrt{45}$
39. $\sqrt{10}\left(\sqrt{40}\right)$
40. $\left(\sqrt{20}\right)\left(\sqrt{80}\right)$

41. $\sqrt{3} \cdot \sqrt{11}$

42. $\sqrt{15} \cdot \sqrt{7}$

43. $\left(\sqrt{3}\right)^2$

44. $\left(\sqrt{17}\right)^2$

45. $\left(-\sqrt{12}\right)^2$

46. $\left(-\sqrt{21}\right)^2$

47. $\left(\sqrt{11}\right)^2 + 13$

48. $21 - \left(\sqrt{8}\right)^2$

49. $5^2 + \left(\sqrt{5}\right)^2$

50. $\left(\sqrt{10}\right)^2 - \left(\sqrt{14}\right)^2$

51. $2\left(\sqrt{7}\right)^2$

52. $\left(\sqrt{8}\right)^2 \left(\sqrt{3}\right)^2$

53. $\dfrac{\sqrt{64}}{\sqrt{4}}$

54. $\dfrac{\sqrt{100}}{\sqrt{25}}$

55. $\sqrt{\dfrac{48}{3}}$

56. $\sqrt{\dfrac{98}{2}}$

57. $\dfrac{\sqrt{81}}{6}$

58. $\dfrac{12}{\sqrt{36}}$

59. $\dfrac{\sqrt{80}}{\sqrt{5}}$

60. $\dfrac{\sqrt{125}}{\sqrt{5}}$

61. $\dfrac{\sqrt{28}}{\sqrt{4}}$

62. $\dfrac{\sqrt{63}}{\sqrt{9}}$

63. $\dfrac{\sqrt{24}}{\sqrt{12}}$

64. $\dfrac{\sqrt{90}}{\sqrt{15}}$

Choose the correct answer.

65. Which of the following is not a perfect square?

(A) 121 (B) $\dfrac{49}{100}$ (C) 0 (D) 0.1 (E) 0.01

66. Which of the following is a perfect square?

(A) 1.6 (B) 196 (C) –36 (D) $\dfrac{56}{81}$ (E) 250

67. Which of the following is true if $n < 4$, where n is a positive real number?

(A) $\sqrt{n} < 2$ (B) $\sqrt{n} > 2$ (C) $\sqrt{n} > 16$ (D) $n^2 > 4^2$ (E) $-\sqrt{n} < -\sqrt{4}$

CHAPTER 3   DECIMALS   PREALGEBRA REVIEW

68. Which of the following is true if $\sqrt{n} > 1$, where n is a positive real number?

   (A) $n < 1$ (B) $n^2 < 1$ (C) $-n > 1$ (D) $n > 1$ (E) $-\sqrt{n} > -1$

69. If $8 < \sqrt{x} < 9$, where x is a positive real number, then

   (A) $8 < x < 9$ (B) $\sqrt{8} < x < 3$ (C) $64 > x > 81$
   (D) $64 < x < 81$ (E) $16 < x < 18$

70. If $9 < x < 16$, where x is a positive real number, then

   (A) $9 < \sqrt{x} < 16$ (B) $81 < \sqrt{x} < 256$ (C) $3 < \sqrt{x} < 4$
   (D) $18 < \sqrt{x} < 32$ (E) $9 > \sqrt{x} > 16$

71. Given the equation $a^2 + b^2 = c^2$, where a, b, and c are positive real numbers, determine the value of c when $a = \sqrt{3}$ and $b = 4$.

   (A) $\sqrt{7}$ (B) 25   (C) 5   (D) 19   (E) $\sqrt{19}$

72. Given the equation $a^2 + b^2 = c^2$, where a, b, and c are positive real numbers, find the value of a when $b = \sqrt{5}$ and $c = 3$.

   (A) 4   (B) $\sqrt{8}$ (C) 2   (D) $\sqrt{14}$ (E) 9

73. The area of a square is 64 square centimeters. What is the length of one side of the square?

   (A) 16 cm   (B) 8 cm   (C) 4 cm   (D) 12 cm   (E) 2 cm

74. In gymnastics, the floor exercise routine is performed on a mat that is 144 square meters. What is the perimeter of the mat?

   (A) 48 m   (B) 12 m   (C) 24 m   (D) 288 m   (E) 36 m

Chapter 4   Basic Concepts of Algebra

Section 1: Variables and Algebraic Expressions

Algebra is a mathematical language in which the general rules and patterns of arithmetic are expressed. In previous sections of this book, letters such as a and x were used to represent numbers. In algebra, a letter or symbol that is used to represent an unknown quantity or number is called a **variable.** Variables together with arithmetic operations are called **variable expressions.** Some examples of variable expressions are

$$9x, \qquad 4 - 3n, \qquad a^2 + b^2, \qquad \text{and} \qquad \frac{x+3}{8}$$

Note that when a number is written next to a variable with no arithmetic symbol between, the operation of multiplication is implied. For instance, $9x$ means $9 \cdot x$. Likewise, the expression $4 - 3n$ can be written as $4 - 3 \cdot n$.

Evaluating Algebraic Expressions

If the value of the variable in an algebraic expression is known, then the expression can be evaluated for that particular value of the variable. That is, the value of the expression can be calculated for the given value of the variable.

> **To Evaluate an Algebraic Expression:**
> Replace the variable with the given value and simplify.

EXAMPLE 1   Find the value of $5x$ if $x = 4$ and if $x = 5$.

SOLUTION   If $x = 4$:
$$5x = 5 \cdot x$$
$$= 5 \cdot 4 \qquad \textit{Replace x with 4.}$$
$$= 20$$

If $x = 5$:
$$5x = 5 \cdot x$$
$$= 5 \cdot 5 \qquad \textit{Replace x with 5.}$$
$$= 25$$

CHAPTER 4 BASIC CONCEPTS OF ALGEBRA

EXAMPLE 2 Evaluate the expression for $n = 7$.

(a) $3n - 8$ (b) $n^2 + 1$ (c) $\dfrac{2n}{n-3}$

SOLUTION

(a) $3n - 8 = 3 \cdot n - 8$
$= 3 \cdot 7 - 8$ *Replace n with 7.*
$= 21 - 8$
$= 13$

(b) $n^2 + 1 = 7^2 + 1$
$= 49 + 1$
$= 50$

(c) $\dfrac{2n}{n-3} = \dfrac{2 \cdot 7}{7-3}$
$= \dfrac{14}{4}$
$= \dfrac{\cancel{2} \cdot 7}{\cancel{2} \cdot 2}$
$= \dfrac{7}{2}$, or $3\dfrac{1}{2}$

EXAMPLE 3 What is the value of $-4p + q$ when $p = 6$ and $q = -9$?

SOLUTION $-4p + q = -4 \cdot 6 + (-9)$
$= -24 + (-9)$
$= -33$

EXAMPLE 4 Evaluate $4t^2 - 8t + 3$ when $t = -1$, $t = 0$, and $t = \dfrac{1}{2}$

SOLUTION When $t = -1$:
$4t^2 - 8t + 3 = 4 \cdot (-1)^2 - 8 \cdot (-1) + 3$
$= 4 \cdot 1 - (-8) + 3$
$= 4 + 8 + 3$
$= 15$

(Example 4 continued)

When $t = 0$:
$$4t^2 - 8t + 3 = 4 \cdot (0)^2 - 8 \cdot 0 + 3$$
$$= 4 \cdot 0 - 0 + 3$$
$$= 0 + 0 + 3$$
$$= 3$$

When $t = \dfrac{1}{2}$:
$$4t^2 - 8t + 3 = 4 \cdot \left(\dfrac{1}{2}\right)^2 - 8 \cdot \dfrac{1}{2} + 3$$
$$= 4 \cdot \dfrac{1}{4} - 4 + 3$$
$$= 1 - 4 + 3$$
$$= 0$$

Simplifying Algebraic Expressions

Algebraic expressions consist of one or more *terms* separated by addition. A **term** is a number, variable, or the product of numbers and variables. For instance, the algebraic expression $2a - \dfrac{1}{3}b + 1$, which can be written as $2a + \left(-\dfrac{1}{3}b\right) + 1$, has three terms: $2a$, $-\dfrac{1}{3}b$, and 1. The terms $2a$ and $-\dfrac{1}{3}b$ are called **variable terms** and 1 is called a **constant term.**

Each variable term in an algebraic expression has a **numerical coefficient**, or **coefficient.** For instance, the coefficient of the term $2a$ is 2 and the coefficient of the term $-\dfrac{1}{3}b$ is $-\dfrac{1}{3}$.

Note that the coefficient of the variable term n is 1, since $1 \cdot n = n$.

CHAPTER 4 BASIC CONCEPTS OF ALGEBRA PREALGEBRA REVIEW

EXAMPLE 5 Identify the terms of $5x - xy - 3y + 11$.

SOLUTION Write the expression in terms of addition. [Recall from Chapter 1 that $a - b = a + (-b)$.]

$$5x - xy - 3y + 11 = 5x + (-xy) + (-3y) + 11$$

The terms are $5x$, $-xy$, $-3y$, and 11.

Some algebraic expression contain terms that have the exact same variable part. These terms are called **like terms.** For instance, $6b$ and $-2b$ are like terms, whereas $7x$ and $9y$ are not like terms.

To simplify an algebraic expression, *combine* like terms by adding their coefficients.

EXAMPLE 6 Simplify.

(a) $3p + 7p$ (b) $9n + 5 - 8n$ (c) $-4a - b + 6a - 10b$

SOLUTION

(a) $3p + 7p = 10p$

(b) $9n + 5 - 8n = 9n + 5 + (-8n)$ *Rewrite subtractions as addition.*

$= 9n + (-8n) + 5$ *Commutative property of addition.*
$= 1n + 5$ *Combine like terms.*
$= n + 5$

(c) $-4a - b + 6a - 10b = -4a + (-b) + 6a + (-10b)$
$= -4a + 6a + (-b) + (-10b)$
$= -4a + 6a + (-1b) + (-10b)$
$= 2a + (-11b)$
$= 2a - 11b$

Suppose the value of p in Example 6(a) is 2. The value of the original expression when $p = 2$ is $3p + 7p = 3 \cdot 2 + 7 \cdot 2 = 6 + 14 = 20$. The value of the simplified expression when $p = 2$ is $10p = 10 \cdot 2 = 20$. Note that the value of each expression is the same. When an algebraic expression is simplified, the result is an **equivalent expression.** That is, the expression has the same value as the original expression when the variable is replaced by a given value.

PREALGEBRA REVIEW   CHAPTER 4   BASIC CONCEPTS OF ALGEBRA

Some algebraic expressions contain parenthesis. To simplify these expressions, use the distributive property to remove the parenthesis and combine any like terms.

Distributive Property: For any numbers a, b, and c, $a(b+c) = a \cdot b + a \cdot c$.

EXAMPLE 7   Simplify.

   (a) $3(2x + 4)$ **(b)** $12 + 2(5t - 1)$ **(c)** $y - 4(2x - y) + 5x$

SOLUTION

(a) $3(2x+4) = 3 \cdot 2x + 3 \cdot 4$ *Apply the distributive property.*
$ = 6x + 12$

(b) $12 + 2(5t - 1) = 12 + 2[5t + (-1)]$ *Rewrite subtractions as additions.*

$ = 12 + 2 \cdot 5t + 2 \cdot (-1)$ *Apply the distributive property.*
$ = 12 + 10t + (-2)$
$ = 12 + (-2) + 10t$
$ = 10 + 10t$ *Combine like terms.*

(c) $y - 4(2x - y) + 5x = y + (-4)[2x + (-1y)] + 5x$
$ = 1y + (-4) \cdot 2x + (-4) \cdot (-1y) + 5x$
$ = 1y + (-8x) + 4y + 5x$
$ = 1y + 4y + (-8x) + 5x$
$ = 5y + (-3x)$
$ = 5y - 3x$

CHAPTER 4   BASIC CONCEPTS OF ALGEBRA

Section 1: Exercises

Evaluate the expression for the given value(s) of the variable(s).

1. $12a$ when $a = 6$
2. $-3n$ when $n = 9$
3. $x - 1$ when $x = 1$
4. $5 - t$ when $t = -5$
5. $2y + 3$ when $y = -2$
6. $4p + 1$ when $p = 7$
7. $9t + 7$ when $t = \dfrac{2}{3}$
8. $6 - 4c$ when $c = \dfrac{3}{4}$
9. $x^2 - 8$ when $x = -3$
10. $2a^2 - 1$ when $a = 2$
11. $4h^2$ when $h = \dfrac{1}{6}$
12. $2 - x^2$ when $x = -\dfrac{1}{3}$
13. $4x - 3y$ when $x = 1, y = 4$
14. $6a + b$ when $a = 2, b = -8$
15. $\dfrac{2x - 1}{x + 5}$ when $x = 6$
16. $\dfrac{9 + n}{3n}$ when $n = -3$
17. $\dfrac{4x}{5y}$ when $x = -0.4, y = 0.8$
18. $\dfrac{a - b}{ab}$ when $a = 1.2, b = 0.6$
19. $s^3 - st + 3t^2$ when $s = -2, t = 4$
20. $2pq - p^2 + q$ when $p = -1, q = -2$

Identify the terms of the expression.

21. $5m + 2n - 3$
22. $-t^2 + \dfrac{1}{2}t - 4st + 11s - 1$

Simplify.

23. $x + 8x$
24. $6n - 10n$
25. $3c - 5 - 2c$
26. $4a + b - a$

PREALGEBRA REVIEW   CHAPTER 4 BASIC CONCEPTS OF ALGEBRA

27. $-2n+5n-9n$

28. $-x+4y-2y+3x$

29. $10a-3b+4b-7a$

30. $13-8t-t-12+4t$

31. $3xy-6x+3y+2xy+7x$

32. $-5b-ab+5b+2ab-a$

33. $2(3x-1)$

34. $-4(x+5)$

35. $6(a+2b)$

36. $4(3m-5n)$

37. $-3(4+3x)+9$

38. $15+2(5x-8)$

39. $1-7x-5(3-2x)$

40. $10y-3(6y-7)+9$

41. $-8p+3(4p-q)-q$

42. $9y-5x-2(2x-6y)$

43. $-4(x-1)+3(3x-5)$

44. $2(p-2q)-3(p+q)$

Choose the correct answer.

45. The expression $2x + 1$ is equivalent to which of the following expressions?

    (A) $3x + 1 + 2x$ **(B)** $4 - x - 2x - 3$ **(C)** $2 + 5x - 1 - 3x$
    (D) $x + 1 - x$ **(E)** $4x - 2x - 2 + 1$

46. Which of the following expressions is not equivalent to $4a - ab + 5b$?

    (A) $b + 4(a + b) - ab$ **(B)** $3a - 2ab + 5b + ab + a$ **(C)** $5(a + b) - ab - a$
    (D) $5b - a + ab + 5a$ **(E)** $5(b - ab) + 4(a + ab)$

47. What is the value of $12t - 3s$ when $t = -\dfrac{2}{3}$ and $s = 2.4$?

    (A) -15.2 **(B)** 0.8 **(C)** 15.2 **(D)** -3.2 **(E)** -0.8

48. $3(xy - 3y + 4) - 4(1 - y) + y - 3xy = ?$

    (A) $-8(y - 1)$ **(B)** $xy - 4(y - 2)$ **(C)** $4(y + 2)$
    (D) $-4(y - 2)$ **(E)** $-4(2 - y)$

CHAPTER 4   BASIC CONCEPTS OF ALGEBRA   PREALGEBRA REVIEW

Section 2: Solving One-Step Equations

An **equation** is a mathematical statement that two quantities or expressions are equal. For example, $2x = 6$ is an equation. A **solution** of an equation is a value that when replaced for the variable in the equation results in a true statement. For instance, 3 is a solution of $2x = 6$ because replacing 3 for x in the equation results in a true statement. That is, $2 \cdot 3 = 6 \rightarrow 6 = 6$.

Two equations with the same solutions are called **equivalent equations.** For example, the equations $2x = 6$ and $x = 3$ are equivalent equations because they both have the same solution, 3.

EXAMPLE 1   Is 7 a solution of $3x - 9 = 12$?

SOLUTION   Replace the 7 for x in the equation and simplify.

$$3x - 9 = 12$$
$$3 \cdot 7 - 9 \stackrel{?}{=} 12$$
$$21 - 9 \stackrel{?}{=} 12$$
$$12 = 12$$

Since $12 = 12$ is a true statement, 7 is a solution of the equation.

Solving Equations of the Form $x + b = c$

To **solve** an equation means to find the value of the variable that makes the equation a true statement. The solution is obtained by writing a series of equivalent equations to reach an equivalent equation of the form

$$x = \text{a number}$$
$\uparrow$
variable (with a coefficient of 1)

That is, an equivalent equation in which the variable is alone on one side of the equation. This is sometimes referred to as isolating the variable.

An equation of the form $x + b = c$, where b and c are numbers, can be solved using the *Addition Property of Equality*.

Addition Property of Equality: For any numbers a, b, and c, if $a = b$, then $a + c = b + c$.

The addition property says that adding the same quantity to both sides of an equation results in an equivalent equation. For instance, given that $4 = 4$, it follows that $4 + 2 = 4 + 2$.

Recall that the sum of a number and its additive inverse, or opposite, is zero. This fact along with the Addition Property of Equality will be used to solve equations of the form $x + b = c$.

EXAMPLE 2 Solve $x + 3 = 7$ for x.

SOLUTION To get x alone one the left side of the equation, add the opposite of 3 to both sides of the equation.

$$x + 3 = 7$$
$$x + 3 + (-3) = 7 + (-3) \qquad \text{Add the opposite of 3 to both sides.}$$
$$x + 0 = 4$$
$$x = 4$$

The solution is 4. Note that replacing 4 for x in the original equation results in true statement.

EXAMPLE 3 Solve for x.

(a) $4 = x - 9$ \qquad (b) $x - 5 = -12$

SOLUTION

(a)
$$4 = x - 9$$
$$4 = x + (-9) \qquad \text{Rewrite the subtraction as addition.}$$
$$4 + 9 = x + (-9) + 9 \qquad \text{Add the opposite of } -9 \text{ to both sides.}$$
$$13 = x + 0$$
$$13 = x$$

(b)
$$x - 5 = -12$$
$$x + (-5) = -12$$
$$x - 5 + 5 = -12 + 5 \qquad \text{Add the opposite of } -5 \text{ to both sides.}$$
$$x + 0 = -7$$
$$x = -7$$

CHAPTER 4 BASIC CONCEPTS OF ALGEBRA

EXAMPLE 4 Solve.

(a) $n + \dfrac{3}{4} = 2$ (b) $-\dfrac{1}{2} = t - \dfrac{2}{3}$

SOLUTION

(a) $$n + \dfrac{3}{4} = 2$$

$$n + \dfrac{3}{4} + \left(-\dfrac{3}{4}\right) = 2 + \left(-\dfrac{3}{4}\right)$$

$$n + 0 = \dfrac{8}{4} + \left(-\dfrac{3}{4}\right)$$

$$n = \dfrac{5}{4}, \text{ or } 1\dfrac{1}{4}$$

(b) $$-\dfrac{1}{2} = t - \dfrac{2}{3}$$

$$-\dfrac{1}{2} + \dfrac{2}{3} = t + \left(-\dfrac{2}{3}\right) + \dfrac{2}{3}$$

$$-\dfrac{3}{6} + \dfrac{4}{6} = t + 0$$

$$\dfrac{1}{6} = t$$

Solving Equations of the Form $ax = c$

An equation of the form $ax = c$, where a and c are numbers, can be solved using the *Division Property of Equality* or the *Multiplication Property of Equality*.

Division Property of Equality: For any numbers a, b, and c, where $c \neq 0$, if $a = b$, then $\dfrac{a}{c} = \dfrac{b}{c}$.

The division property says that dividing both sides of an equation by the same nonzero number results in an equivalent equation. For instance, given that $4 = 4$, it follows that $\dfrac{4}{2} = \dfrac{4}{2}$.

Multiplication Property of Equality: For any numbers a, b, and c, where $c \neq 0$, if $a = b$, then $a \cdot c = b \cdot c$.

PREALGEBRA REVIEW   CHAPTER 4 BASIC CONCEPTS OF ALGEBRA

The Multiplication Property says that multiplying both sides of an equation by the same nonzero number results in an equivalent equation. For instance, given that $4 = 4$, it follows that $4 \cdot 2 = 4 \cdot 2$.

Recall that when a number is divided by itself the result is 1. This fact along with the Division Property of Equality can be used to solve equations of the form $ax = c$, where a and c are integers or decimals.

EXAMPLE 5   Solve $2x = 16$ for x.

SOLUTION   To get x alone on the left side of the equation, divide both sides by 2, the coefficient of x.

$$2x = 16$$
$$\frac{2x}{2} = \frac{16}{2}$$
$$1 \cdot x = 8$$
$$x = 8$$

The solution is 8. Note that replacing 8 for x in the original equation results in a true statement.

EXAMPLE 6   Solve: **(a)** $-n = 5$ **(b)** $4y = -6$

SOLUTION

(a) $-n = 5$
$$-1 \cdot n = 5$$
$$\frac{-1 \cdot n}{-1} = \frac{5}{-1}$$
$$1 \cdot n = -5$$
$$n = -5$$

(b) $4y = -6$
$$\frac{4y}{4} = \frac{-6}{4}$$
$$1 \cdot y = -\frac{\cancel{2} \cdot 3}{\cancel{2} \cdot 2}$$
$$y = -\frac{3}{2}, \text{ or } -1\frac{1}{2}$$

CHAPTER 4 BASIC CONCEPTS OF ALGEBRA

EXAMPLE 7   Solve: $0.8t = 4$

SOLUTION

$$0.8t = 4$$
$$\frac{0.8t}{0.8} = \frac{4}{0.8}$$
$$1 \cdot t = 5$$
$$t = 5$$

Recall that when a number is multiplied by its reciprocal the result is 1. This fact along with the Multiplication Property of Equality can be used to solve equations of the form $ax = c$, where a or c is a fraction.

EXAMPLE 8   Solve: **(a)** $\frac{2}{5}x = \frac{2}{9}$ **(b)** $4p = \frac{1}{2}$

SOLUTION

(a) To get x alone on the left side of the equation, divide both sides by $\frac{5}{2}$, the reciprocal of the coefficient of x.

$$\frac{2}{5}x = \frac{2}{9}$$
$$\frac{5}{2} \cdot \frac{2}{5}x = \frac{5}{2} \cdot \frac{2}{9}$$
$$1 \cdot x = \frac{5 \cdot \cancel{2}}{\cancel{2} \cdot 9}$$
$$x = \frac{5}{9}$$

(b) $4p = \frac{1}{2}$
$$\frac{1}{4} \cdot \left(\frac{4}{1}p\right) = \frac{1}{4} \cdot \frac{1}{2}$$
$$1 \cdot p = \frac{1}{8}$$
$$p = \frac{1}{8}$$

EXAMPLE 9 Solve.

(a) $-1 = \dfrac{a}{-6}$ (b) $\dfrac{3}{8}n = 3\dfrac{1}{2}$

SOLUTION

(a) $$-1 = \dfrac{a}{-6}$$
$$-1 = \dfrac{1}{-6}a$$
$$\dfrac{-6}{1} \cdot (-1) = \dfrac{-6}{1} \cdot \left(\dfrac{1}{-6}\right) a$$
$$6 = 1 \cdot a$$
$$6 = a$$

(b) $$\dfrac{3}{8}n = 3\dfrac{1}{2}$$
$$\dfrac{3}{8}n = \dfrac{7}{2}$$
$$\dfrac{8}{3} \cdot \dfrac{3}{8}n = \dfrac{8}{3} \cdot \dfrac{7}{2}$$
$$1 \cdot n = \dfrac{\overset{1}{\cancel{2}} \cdot 2 \cdot 2 \cdot 7}{3 \cdot \underset{1}{\cancel{2}}}$$
$$1 \cdot n = \dfrac{28}{3}$$
$$n = \dfrac{28}{3}, \text{ or } 9\dfrac{1}{3}$$

CHAPTER 4   BASIC CONCEPTS OF ALGEBRA

PREALGEBRA REVIEW

Section 2: Exercises

Answer the question.

1. Is 6 a solution of the equation $3x - 5 = 13$?

2. Is $\frac{1}{2}$ a solution of the equation $7 - 2n = 8$?

Solve for the variable.

3. $n + 5 = 11$
4. $x - 4 = 8$
5. $t - 11 = 1$
6. $y + 12 = 12$
7. $p + 8 = 3$
8. $a + 6 = 5$
9. $c - 1 = -3$
10. $-2 = n - 10$
11. $-15 = x + 7$
12. $-11 + b = -21$
13. $4y = 24$
14. $8n = 56$
15. $3x = -27$
16. $-2p = -32$
17. $6t = \frac{2}{3}$
18. $\frac{5}{8} = -5a$
19. $10b = 8$
20. $-9x = -6$
21. $\frac{3}{4}w = 12$
22. $\frac{5}{6}x = 20$
23. $\frac{4}{15} = \frac{8}{9}n$
24. $\frac{2}{3}q = -\frac{6}{7}$
25. $\frac{n}{5} = -5$
26. $-7 = \frac{x}{-4}$
27. $-\frac{4}{5}a = 2\frac{1}{10}$
28. $-3\frac{1}{3}x = \frac{5}{12}$
29. $t - \frac{1}{4} = 1$
30. $x + \frac{2}{5} = -3$
31. $p + \frac{1}{6} = 1\frac{2}{3}$
32. $y - \frac{5}{8} = -2\frac{1}{12}$
33. $2.8 + x = 4.9$
34. $a - 5.3 = -1.1$
35. $0.4n = 2$
36. $-2t = 6.3$
37. $-2.4p = -2.88$
38. $5.75 = 4.6y$

PREALGEBRA REVIEW   CHAPTER 4 BASIC CONCEPTS OF ALGEBRA

Section 3: Solving Two-Step Equations

An equation of the form $ax + b = c$ can be solved using the *Addition Property of Equality* and either the *Multiplication* or *Division Property of Equality*.

To solve equations in this form, first use the Addition Property to isolate the variable term on one side of the equation. Then use the Multiplication or Division Property to obtain a variable term with a coefficient of 1.

EXAMPLE 1   Solve: $2x + 5 = 9$

SOLUTION
$$2x + 5 = 9$$
$$2x + 5 + (-5) = 9 + (-5)$$
$$2x = 4$$
$$\frac{2x}{2} = \frac{4}{2}$$
$$x = 2$$

EXAMPLE 2   Solve for the variable.

(a) $1 - 3y = -11$ (b) $-x + 8 = 4$

SOLUTION

(a)
$$1 - 3y = -11$$
$$1 + (-3y) = -11$$
$$-1 + 1 + (-3y) = -11 + (-1)$$
$$\frac{-3y}{-3} = \frac{-12}{-3}$$
$$y = 4$$

(b)
$$-x + 8 = 4$$
$$-1 \cdot x + 8 = 4$$
$$-1 \cdot x + 8 + (-8) = 4 + (-8)$$
$$-1 \cdot x = -4$$
$$\frac{-1 \cdot x}{-1} = \frac{-4}{-1}$$
$$x = 4$$

CHAPTER 4 BASIC CONCEPTS OF ALGEBRA

EXAMPLE 3 Solve: $\dfrac{2}{3}t - 1 = 3$

SOLUTION
$$\dfrac{2}{3}t - 1 = 3$$
$$\dfrac{2}{3}t + (-1) = 3$$
$$\dfrac{2}{3}t + (-1) + 1 = 3 + 1$$
$$\dfrac{2}{3}t = 4$$
$$\dfrac{3}{2} \cdot \dfrac{2}{3}t = \dfrac{3}{2} \cdot \dfrac{4}{1}$$
$$t = \dfrac{3 \cdot \cancel{2} \cdot 2}{\cancel{2} \cdot 1}$$
$$t = \dfrac{6}{1} = 6$$

EXAMPLE 4 Solve: $-\dfrac{5}{9}n - \dfrac{1}{6} = 2$

SOLUTION
$$-\dfrac{5}{9}n - \dfrac{1}{6} = 2$$
$$-\dfrac{5}{9}n + \left(-\dfrac{1}{6}\right) = 2$$
$$-\dfrac{5}{9}n + \left(-\dfrac{1}{6}\right) + \left(\dfrac{1}{6}\right) = 2 + \dfrac{1}{6}$$
$$-\dfrac{5}{9}n = 2\dfrac{1}{6}$$
$$-\dfrac{9}{5} \cdot \left(-\dfrac{5}{9}n\right) = -\dfrac{9}{5} \cdot \dfrac{13}{6}$$
$$n = -\dfrac{\cancel{3} \cdot 3 \cdot 13}{5 \cdot 2 \cdot \cancel{3}}$$
$$n = -\dfrac{39}{10}, \text{ or } -3\dfrac{9}{10}$$

PREALGEBRA REVIEW

Section 3: Exercises

1. $3n - 2 = 10$

2. $7x + 1 = 8$

3. $-4a + 9 = 25$

4. $5t - 3 = -13$

5. $2t - 7 = -4$

6. $-3y - 10 = -5$

7. $2 - x = -6$

8. $9 = 5 - 4p$

9. $\dfrac{b}{4} + 9 = 8$

10. $\dfrac{y}{2} - 11 = -7$

11. $-\dfrac{3}{5}n - 3 = 15$

12. $\dfrac{3}{8}a + 12 = -9$

13. $5y - 2 = \dfrac{1}{7}$

14. $4n + \dfrac{5}{8} = 1$

15. $\dfrac{1}{8} - \dfrac{3}{4}x = -1$

16. $-\dfrac{3}{5}y + 2 = -\dfrac{1}{10}$

17. $8t - 5.2 = -2$

18. $1.9 - 5x = 0.4$

19. $2.4p + 7.2 = -3.6$

20. $-1.44 - 0.8n = 6.4$

Section 4: Algebra and Problem Solving

In addition to providing a way to represent general rules and patterns of arithmetic, algebra provides a way to model and solve applied problems in which one or more quantities are unknown. The key to using algebra as a problem solving tool is in translating word phrases into algebraic expressions and in translating word sentences into algebraic equations.

Translating Word Phrases into Algebraic Expressions

Recall from Chapter 1 the key words and phrases associated with arithmetic operations that sometimes appear in applied problems.

Addition	Subtraction	Multiplication	Division
add	subtract	multiply	divide
sum	difference	product	quotient
plus	minus	times	divided by
more than	less than	multiplied by	
total	decreased by	of	
added to	subtracted from		
increased by			

EXAMPLE 1   Translate the word phrase to an algebraic expression. Use n as the variable for the unknown number.

(a) 9 more than a number
(b) The sum of a number and 15
(c) A number subtracted from 8
(d) 4 less than a number

SOLUTION
(a) $n + 9$
(b) $n + 15$
(c) $8 - n$
(d) $n - 4$

Note that because of the Commutative Property of Addition, the expression in Example 1(a) can also be written as $9 + n$. Likewise, the expression in Example 1(b) can also be written as $15 + n$.

PREALGEBRA REVIEW — CHAPTER 4 BASIC CONCEPTS OF ALGEBRA

EXAMPLE 2 Translate the word phrase to an algebraic expression. Use x as the variable for the unknown number.

(a) 5 multiplied by a number
(b) One-third of a number
(c) 4 divided by a number
(d) The quotient of a number and 21

SOLUTION

(a) $5 \cdot x$, or $5x$

(b) $\frac{1}{3} \cdot x$, or $\frac{1}{3}x$

(c) $4 \div x$, or $\frac{4}{x}$

(d) $x \div 21$, or $\frac{x}{21}$

Note that because of the Commutative Property of Multiplication, the expression in Example 2(a) can also be written as $x \cdot 5$. Likewise, the expression in Example 2(b) can also be written as $x \cdot \frac{1}{3}$. In algebra, however, the constant factor is generally written in front of the variable.

In many problem situations, a word phrases will contain more than one operation.

EXAMPLE 3 Translate the word phrase to an algebraic expression. Use n as the variable for the unknown number.

(a) Twice a number increased by 3
(b) 11 less than one-half of a number
(c) 7 times the sum of a number and 1
(d) The quotient of a number and 5 subtracted from 10

SOLUTION

(a) $2n + 3$

(b) $\frac{1}{2}n - 11$

(c) The sum of a number and 1 is $n + 1$. Seven times the sum is $7(n + 1)$. Note that parentheses are used around the sum.

(d) $10 - \frac{n}{5}$

CHAPTER 4   BASIC CONCEPTS OF ALGEBRA

Translating Word Sentences into Algebraic Equations

The main difference between an algebraic expression and an algebraic equation is the inclusion of the = symbol. Recall that an equation is a mathematical statement that two quantities or expressions are equal.

In translating word statements into algebraic expressions, the following key words will translate to an = sign: equals, is, is equal to, and the result is.

EXAMPLE 4   Translate the word sentence to an equation. Use x as the variable for the unknown quantity.

(a) The sum of a number and 8 is 17
(b) Twice a number divided by 3 equals 1
(c) If 9 is decreased by 4 times a number, the result is 2
(d) 15 more than a number divided by 6 is equal to 7

SOLUTION

(a) $x + 8 = 17$
(b) $\dfrac{2x}{3} = 1$
(c) $9 - 4x = 2$
(d) $\dfrac{x}{6} + 15 = 7$

Section 4: Exercises

Translate the word phrase into an algebraic expression. Use n as the variable for the unknown number.

1. A number subtracted from 9
2. A number times 12
3. A number increased by 20
4. The sum of 4 and a number
5. 16 divided by a number
6. Three-fourths of a number
7. The difference of a number and 1
8. The quotient of a number and 8
9. The product of a number and 5 decreased by 2
10. 6 plus the quotient of a number and 3
11. One-half the sum of a number and 13
12. 4 times the difference between 7 and a number

Translate the word sentence into an algebraic equation. Use x as the variable for the unknown number.

13. A number decreased by 14 is equal to 7
14. 18 is two-thirds of a number
15. 5 subtracted from 3 times a number is 10
16. If a number divided by 8 is increased by 1, the result is 9
17. 16 is equal to 2 times the sum of a number and 12
18. The difference of a number and 6, divided by 30 equals 8
19. 5 less than 9 times a number is equal to 14
20. 25 more than the product of a number and −4 is 2

CHAPTER 4   BASIC CONCEPTS OF ALGEBRA

PREALGEBRA REVIEW

Choose the correct answer.

21. Jay makes $11.25 for each hour he works. Write an algebraic expression that represents the amount he earns for working h hours.

   (A) $11.25 - h$ (B) $\dfrac{11.25}{h}$ (C) $11.25h$ (D) $h + 11.25$ (E) $11.25 + 11.25h$

22. On an exam, a student loses 3 points for each incorrect answer. If a student can earn a maximum of 120 points on the exam, write an algebraic expression that represents his score if he answers n questions incorrectly.

   (A) $\dfrac{120}{3n}$ (B) $120 - 3n$ (C) $120 - n$ (D) $3n - 120$ (E) $120 + 3n$

23. The length of a rectangle is 1 more than twice its width, w. Write an algebraic expression that represents the length of the rectangle.

   (A) $3 + w$ (B) $2w - 1$ (C) $w + 2$ (D) $\dfrac{w}{2} + 1$ (E) $2w + 1$

24. Pedro drove a total of 192 miles at a constant speed of 64 miles per hour. Which equation can be used to find the number of hours, h, he drove?

   (A) $\dfrac{64}{h} = 192$ (B) $64h = 192$ (C) $64 + h = 192$

   (D) $\dfrac{h}{192} = 64$ (E) $192 - h = 64$

25. A fitness center charges an $40 per month plus a one time enrollment fee of $25. Which equation can be used to determine the number of months, m, a member has been enrolled if she pays a total of $305 for her membership?

   (A) $40m + 25 = 305$ (B) $25m + 40 = 305$ (C) $25m + 40 = 305$
   (D) $40m - 25 = 305$ (E) $305 - 25m = 40$

26. Gina sells necklaces for $30.00 each. It costs her $15.85 in supplies to make one necklace. Which equation can be used to find the number of necklaces, n, she must sell in order to make a profit of $566.00?

   (A) $15.85n = 566$ (B) $14.15n = 566$ (C) $15.85n + 30 = 566$
   (D) $15.85 - 30n = 566$ (E) $30n + 15.85 = 566$

Chapter 5 Ratios, Rates, and Proportions

Section 1: Ratios

A **ratio** compares two quantities with the *same* units. The ratio of the quantity a to the quantity b can be written in three ways:

$$a \text{ to } b \quad \text{or} \quad a:b \quad \text{or} \quad \frac{a}{b}$$

For instance, the ratio of 5 days to 7 days can be written as 5 to 7 or 5 : 7 or $\frac{5}{7}$.

Note that the colon and the fraction bar are read as "to." Also, note that units are not written in a ratio.

Most of the time, ratios are written as fractions. The first quantity mentioned is the numerator and the second quantity mentioned is the denominator. When a ratio is written as a fraction, it should be written in lowest terms.

EXAMPLE 1 Express the ratio as a fraction.

    (a) 20 milligrams of medicine to 50 milligrams of medicine
    (b) 12 months to 4 months

SOLUTION

    (a) $\dfrac{20}{50} = \dfrac{\cancel{10} \cdot 2}{\cancel{10} \cdot 5} = \dfrac{2}{5}$

So, the ratio of 20 milligrams of medicine to 50 milligrams of medicine is $\dfrac{2}{5}$, or 2 to 5.

    (b) $\dfrac{12}{4} = \dfrac{\cancel{4} \cdot 3}{\cancel{4} \cdot 1} = \dfrac{3}{1}$

So, the ratio of 12 months to 4 months is $\dfrac{3}{1}$, or 3 to 1.

EXAMPLE 2 Express the ratio as a fraction.

(a) 45 inches to 3 feet
(b) 5 hours to 90 minutes

SOLUTION Remember that a ratio compares two quantities with the *same* units.

(a) First convert 3 feet to inches. Since 1 foot = 12 inches, 3 feet is $3 \cdot 12 = 36$ inches.

Next, write a ratio: $\dfrac{45}{36}$

Now, simplify: $\dfrac{45}{36} = \dfrac{\cancel{9} \cdot 5}{\cancel{9} \cdot 4} = \dfrac{5}{4}$

(b) First convert 5 hours to minutes. Since 1 hour = 60 minutes, 5 hours is $5 \cdot 60 = 300$ minutes.

Next, write a ratio: $\dfrac{300}{90}$

Now, simplify: $\dfrac{300}{90} = \dfrac{\cancel{30} \cdot 10}{\cancel{30} \cdot 3} = \dfrac{10}{3}$

When a ratio contains decimals, multiply the numerator and denominator by the appropriate power of 10 to rewrite the fraction as the ratio of two whole numbers. Then simplify the ratio, if possible.

EXAMPLE 3 Write the ratio of 1.75 liters to 3.25 liters as a fraction.

SOLUTION First, write the ratio. Since the decimals in numerator and denominator have two decimal places, multiply the numerator and denominator by 100.

$$\dfrac{1.75}{3.25} = \dfrac{1.75 \cdot 100}{3.25 \cdot 100} = \dfrac{175}{325}$$

Now simplify: $\dfrac{175}{325} = \dfrac{\cancel{25} \cdot 7}{\cancel{25} \cdot 13} = \dfrac{7}{13}$

EXAMPLE 4 Write the ratio of 3 meters to $4\frac{1}{2}$ meters as a fraction.

SOLUTION First, write the ratio. Since the denominator contains a mixed number, rewrite denominator as an improper fraction.

$$\frac{3}{4\frac{1}{2}} = \frac{3}{\frac{9}{2}}$$

Now divide the numerator by the denominator and simplify.

$$\frac{3}{\frac{9}{2}} = 3 \div \frac{9}{2} = \frac{3}{1} \div \frac{9}{2} = \frac{3}{1} \cdot \frac{2}{9} = \frac{\cancel{3} \cdot 2}{1 \cdot \cancel{9} \cdot 3} = \frac{2}{3}$$

Solving Applied Problems

EXAMPLE 5 In a company survey, the Human Resources department found that 5 of every 8 employees were satisfied with their benefits package. What is the ratio of the number of employees satisfied with their benefits package to the total number of employees?

SOLUTION The number of employees satisfied with their benefits package and the total number of employees is given. The key phrase here is "5 of every 8." The ratio is:

$$\frac{\text{number of employees satisfied}}{\text{total number of employees}} = \frac{5}{8}$$

Note that the ratio is in simplest form.

EXAMPLE 6 At a local service station, the price of a gallon of regular gasoline increased from $1.85 to $2.15. Find the ratio of the increase in price to the original price.

SOLUTION The original price and the new price are given. To find the increase in price, subtract.

New price − Original price   → $2.15 − $1.85

$2.15 − 1.85 = 0.30$

The increase in price is $0.30. The ratio of the increase in price to the original price is:

$$\frac{\text{increase in price}}{\text{original price}} = \frac{0.30}{1.85}$$

Simplify the ratio: $\dfrac{0.30 \cdot 100}{1.85 \cdot 100} = \dfrac{30}{185} = \dfrac{\cancel{5} \cdot 6}{\cancel{5} \cdot 37} = \dfrac{6}{37}$

PREALGEBRA REVIEW   CHAPTER 5   RATIOS, RATES, AND PROPOTIONS

Section 1: Exercises

Write the ratio in three ways. Do not simplify.

1. 2 weeks to 3 weeks
2. 15 milligrams to 30 milligrams
3. 7 women out of 10 women
4. 1 dollar of every 5 dollars

Express each ratio as a fraction in lowest terms.

5. $6 : 8$
6. $12 : 9$
7. 3 years to 5 years
8. 1 employee out of 4 employees
9. 1.8 miles to 3.6 miles
10. $2.48 to $0.88
11. $4\frac{1}{4}$ hours to 8 hours
12. 1 cup to $\frac{2}{3}$ cup
13. 12 days out of 3 weeks
14. 15 months out of 2 years
15. 4 pounds to 80 ounces
16. 10 yards to 36 feet

Choose the correct answer.

17. What is the ratio $9 : 0.81$ expressed as a fraction in simplest form?

    (A) $\dfrac{9}{0.81}$ (B) $\dfrac{100}{9}$ (C) $\dfrac{10}{9}$ (D) $\dfrac{900}{81}$ (E) $\dfrac{1}{9}$

18. What is the ratio $3\frac{1}{6}$ to $5\frac{3}{4}$ expressed as a fraction in simplest form?

    (A) $\dfrac{38}{69}$ (B) $\dfrac{437}{24}$ (C) $\dfrac{19}{23}$ (D) $\dfrac{76}{138}$ (E) $\dfrac{2}{3}$

19. Express 0.5 mile to 1320 feet as a ratio.

    (A) $1 : 2$ (B) $2 : 1$ (C) $1 : 4$ (D) $4 : 1$ (E) $1 : 6$

CHAPTER 5 RATIOS, RATES, AND PROPORTIONS   PREALGEBRA REVIEW

20. Express the ratio of 384 meters to 4.8 kilometers as a fraction in lowest terms.

   (A) $\dfrac{80}{1}$ (B) $\dfrac{4}{5}$ (C) $\dfrac{8}{1}$ (D) $\dfrac{2}{25}$ (E) $\dfrac{2}{50}$

21. In a taste test, 3 out of 5 people preferred Brand A over Brand B. Write a ratio that shows the number of people who preferred Brand B out of the total number of people.

   (A) $\dfrac{3}{5}$ (B) $\dfrac{1}{4}$ (C) $\dfrac{2}{5}$ (D) $\dfrac{2}{3}$ (E) $\dfrac{3}{8}$

22. The Wallaces put a $45,000 down payment on a $225,000 home. What is the ratio of the down payment to the price of the home?

   (A) $\dfrac{9}{45}$ (B) $\dfrac{5}{1}$ (C) $\dfrac{1}{5}$ (D) $\dfrac{45}{225}$ (E) $\dfrac{3}{15}$

23. Dee gets 45 minutes for lunch in an 8 hour shift. What is the ratio of the time she gets for lunch to the time of her shift?

   (A) $\dfrac{3}{2}$ (B) $\dfrac{45}{8}$ (C) $\dfrac{15}{1600}$ (D) $\dfrac{15}{16}$ (E) $\dfrac{3}{32}$

24. There are 18 men and 16 women enrolled in a math class. What is the ratio of men enrolled to the total number of people enrolled in the class?

   (A) $\dfrac{9}{17}$ (B) $\dfrac{9}{8}$ (C) $\dfrac{8}{17}$ (D) $\dfrac{9}{34}$ (E) $\dfrac{17}{9}$

25. Last year, Cam's car insurance premium was $552. This year his insurance premium decreased by $84. Write the ratio of Cam's car insurance premium this year to his car insurance premium last year.

   (A) $\dfrac{7}{46}$ (B) $\dfrac{53}{46}$ (C) $\dfrac{39}{46}$ (D) $\dfrac{7}{39}$ (E) $\dfrac{46}{39}$

26. A college increased the cost of one credit from $120 to $145. Write the ratio of the increase in cost for a one credit to the original cost of one credit.

   (A) $\dfrac{5}{24}$ (B) $\dfrac{24}{29}$ (C) $\dfrac{5}{29}$ (D) $\dfrac{53}{24}$ (E) $\dfrac{29}{24}$

PREALGEBRA REVIEW   CHAPTER 5 RATIOS, RATES, AND PROPOTIONS

Section 2: Rates

A **rate** compares two quantities with different units. A rate can be written as a fraction in simplest form. Note that units are included when writing a rate.

EXAMPLE 1   A sales associate earned $54 in 6 hours. Find the earnings to time rate.

SOLUTION

$$\frac{54 \text{ dollars}}{6 \text{ hours}} = \frac{9 \text{ dollars}}{1 \text{ hour}}$$

The rate is $9 for each hour worked or $9 per hour.

In a rate, the quantities being compared are usually separated by one of the following words:

        in   for   on   per   from

EXAMPLE 2   Express the rate as a fraction.

    (a)   351 miles on 13 gallons of gasoline
    (b) $15 for 6 boxes of cookies
    (c)   950 revolutions in 8 minutes

SOLUTION

(a) $\dfrac{351 \text{ miles}}{13 \text{ gallons}} = \dfrac{27 \text{ miles}}{1 \text{ gallon}}$

(b) $\dfrac{15 \text{ dollars}}{6 \text{ boxes}} = \dfrac{5 \text{ dollars}}{2 \text{ boxes}}$

(c) $\dfrac{950 \text{ revolutions}}{8 \text{ minutes}} = \dfrac{475 \text{ revolutions}}{4 \text{ minutes}}$

Unit Rates

A **unit rate** is a rate with a denominator of 1. Some frequently used unit rates include *rate of speed*, which is distance traveled for one unit of time, such as miles per hour and meters per second, and *unit price*, which is the price for one unit, such as dollars per pound.

To find a unit rate, write the rate in fraction form and then divide.

CHAPTER 5   RATIOS, RATES, AND PROPORTIONS   PREALGEBRA REVIEW

EXAMPLE 3   Find the unit rate.

(a) 192 feet in 6 seconds
(b) 240 in 20 cases
(c) $105 for 12 tickets

SOLUTION   Begin by writing each rate in fraction form. Then divide both the numerator and the denominator by the number of units in the denominator.

(a) $\dfrac{192 \text{ feet}}{6 \text{ seconds}} = \dfrac{192 \text{ feet} \div 6}{6 \text{ seconds} \div 6} = \dfrac{32 \text{ feet}}{1 \text{ second}}$, or 32 feet per second

(b) $\dfrac{240 \text{ cans}}{20 \text{ cases}} = \dfrac{240 \text{ cans} \div 20}{20 \text{ cases} \div 20} = \dfrac{12 \text{ cans}}{1 \text{ case}}$, or 12 cans per case

(c) $\dfrac{\$105}{12 \text{ tickets}} = \dfrac{\$105 \div 12}{12 \text{ tickets} \div 12} = \dfrac{\$8.75}{1 \text{ ticket}}$, or $8.75 per ticket

EXAMPLE 4   A grocery store sells a 5 pound bag of potatoes for $3.49. What is the unit price of the potatoes?

SOLUTION   In this situation, the unit price is the price per pound.

$$\dfrac{\$3.49}{5 \text{ pounds}} = \$0.698 \text{ per pound}$$

EXAMPLE 5   Which is a better buy: a 15 ounce box of cereal for $2.50, or a 20 ounce box of the same cereal for $3.99? Round answers to the nearest tenth of a cent.

SOLUTION   The unit rate for the 15 ounce box of cereal is

$$\dfrac{\$2.50}{15 \text{ ounces}} = \dfrac{250 \text{ cents}}{15 \text{ ounces}} \approx 16.7 \text{ cents per ounce}$$

The unit rate for the 20 ounce box of cereal is

$$\dfrac{\$3.89}{20 \text{ ounces}} = \dfrac{389 \text{ cents}}{20 \text{ ounces}} \approx 19.5 \text{ cents per ounce}$$

The unit rate for the 15 ounce box is less than the unit rate for the 20 ounce box, so the 15 ounce box is the better buy.

PREALGEBRA REVIEW   CHAPTER 5 RATIOS, RATES, AND PROPOTIONS

EXAMPLE 6   One jet flies 840 miles in 2 hours. Another jet flies 1245 miles in 3 hours. Which jet is flying at a faster rate?

SOLUTION   The first jet is flying at a rate of $\dfrac{840 \text{ miles}}{2 \text{ hours}} = 420$ miles per hour. The second jet is flying at a rate of $\dfrac{1245 \text{ miles}}{3 \text{ hours}} = 415$ miles per hour. So the first jet is flying at a faster rate.

EXAMPLE 7   A car can travel 329 miles on $\tfrac{3}{4}$ tank of gas. If the tank holds 14.4 gallons of gas, how many miles can the car travel on one gallon of gas? Round to the nearest mile.

SOLUTION   Begin by finding the number of gallons of gas in $\tfrac{3}{4}$ tank. Since the tank holds a total of 14.4 gallons, multiply 14.4 by $\tfrac{3}{4}$.

$$14.4 \cdot \frac{3}{4} = \frac{14.4}{1} \cdot \frac{3}{4} = \frac{14.4 \cdot 3}{4} = \frac{43.2}{4} = 10.8$$

The car can travel 329 miles on 10.8 gallons of gas. Therefore, the car can travel $\dfrac{329 \text{ miles}}{10.8 \text{ gallons}} \approx 30$ miles per gallon.

EXAMPLE 8   Saundra jogs 2 miles in 16 minutes. How fast is she jogging in miles per hour?

SOLUTION   First find the unit rate: $\dfrac{2 \text{ miles}}{16 \text{ minutes}} = \dfrac{0.125 \text{ mile}}{1 \text{ minute}}$, or 0.125 mile per minute

To find the number of miles Saundra jogs in 1 hour, or 60 minutes, multiply the unit rate by 60 minutes:

$$\frac{0.125 \text{ mile}}{\cancel{\text{minute}}} \times 60 \ \cancel{\text{minutes}} = 0.125 \text{ mile} \times 60 = 7.5 \text{ miles}$$

So, Saundra jogs 7.5 miles in 1 hour. That is, she is jogging at a rate of 7.5 miles per hour.

EXAMPLE 9 Hank earns d dollars is h hours of work. Write a variable expression for the amount he earns in t hours.

SOLUTION First determine the unit rate in dollars per hour: $\dfrac{d \text{ dollars}}{h \text{ hours}}$

To find amount he earns in t hours, multiply the amount he earns in one hour (i.e. the unit rate) by t hours:

$$t \text{ \cancel{hours}} \cdot \frac{d \text{ dollars}}{h \text{ \cancel{hours}}} = t \cdot \frac{d}{h} \text{ dollars} = \frac{td}{h} \text{ dollars}.$$

So, a variable expression for the amount Hank earns in t hours is $\dfrac{td}{h}$.

PREALGEBRA REVIEW   CHAPTER 5   RATIOS, RATES, AND PROPOTIONS

Section 2: Exercises

Write each rate as a fraction in lowest terms.

1. 64 meters in 2 seconds
2. 48 cans in 4 boxes
3. 100 miles in 3 hours
4. 6 ounces in 1 serving
5. $20 for 16 dozen
6. 5 buses for 230 people
7. 9 instructors for 165 students
8. $164 in 8 hours
9. 36 cookies for 12 children
10. 462 miles on 14 gallons

Find the unit rate.

11. 300 milligrams in 12 pills
12. 9216 people in 4 square miles
13. 96 hours in 24 days
14. 60 minutes for 3 periods
15. 10 yards for 4 downs
16. 642 bushels from 8 acres
17. $6.98 for 2 pounds
18. $30.10 for 14 gallons
19. 44 yards for 8 dresses
20. $564 for 6 people

Find better buy.

21. Coffee:
    a 12 ounce cup for $1.29
    a 24 ounce cup for $2.49

22. Dog food (bag):
    5 pounds for $8.49
    20 pounds for $22.99

23. Canned peaches:
    8 ounces for $0.59
    14 ounces for $0.79
    28 ounces for $1.59

24. Fabric softener sheets:
    40 sheets for $2.49
    80 sheets for $4.99
    120 sheets for $6.49

Solve.

25. Shar jogs 9 miles in 1.5 hours. Ted jogs 5 miles in 40 minutes. Who is jogging at a faster rate?

CHAPTER 5 RATIOS, RATES, AND PROPORTIONS   PREALGEBRA REVIEW

26. An SUV can travel 184 miles on 8 gallons of gas. A mid-size car can travel 140 miles on 5 gallons of gas. Which vehicle gets more miles per gallon?

Choose the correct answer.

27. An online retailer sells a package of 16 batteries for $8.59. What is the unit price of the batteries?

   (A) $0.50 per battery   (B) $0.55 per battery   (C) $0.54 per battery
   (D) $0.535 per battery (E) $0.53 per battery

28. A home improvement store sells a 5 gallon container of paint for $69.95. Find the unit price of the paint.

   (A) $14.00/gallon   (B) $13.99/gallon   (C) $13.89/gallon
   (D) $13.995/gallon (E) $13.00/gallon

29. An airplane can fly 270 miles in $\frac{2}{3}$ hour. How many miles can the airplane fly per hour?

   (A) 135   (B) 540   (C) 405   (D) 180   (E) 415

30. An administrative assistant is paid $494 for 40 hours of work. How much is she paid per hour?

   (A) $123.50   (B) $12.25   (C) $61.75   (D) $98.80   (E) $12.35

31. A patient's heart beats about 18 times in 15 seconds. How many times does it beat per minute?

   (A) 1.2   (B) 60   (C) 50   (D) 72   (E) 36

32. Sound travels 99,000 feet in 1.5 minutes. At what rate does sound travel in feet per second?

   (A) 66,000   (B) 1100   (C) 660   (D) 11,000   (E) 1650

33. A lawn mower can mow 4 acres on $\frac{1}{2}$ tank of gas. If the tank holds 7.2 gallons of gas, how many acres can the mower mow on one gallon of gas? Round to the nearest tenth of an acre.

   (A) 2.0   (B) 3.6   (C) 1.1   (D) 1.8   (E) 0.9

34. A contractor can plaster 75 square feet of wall from $\frac{3}{4}$ pound of plaster. If a bag contains 50 pounds of plaster, how many square feet are plastered from one pound of plaster?

　　(A) 12.5　　(B) 2　　(C) 37.5　　(D) 25　　(E) 5

35. A nurse works h hours in d days. Write the number of hours she works in t days as an algebraic expression.

　　(A) $\dfrac{d}{h}$　　(B) $\dfrac{th}{d}$　　(C) $\dfrac{t}{d}$　　(D) $\dfrac{t}{dh}$　　(E) $\dfrac{d}{th}$

36. A manufacturer ships a units in x days. Express the number of units the manufacturer ships in b days as an algebraic expression.

　　(A) $\dfrac{a}{x}$　　(B) $\dfrac{x}{ab}$　　(C) $\dfrac{xa}{b}$　　(D) $\dfrac{ab}{x}$　　(E) $\dfrac{b}{x}$

CHAPTER 5   RATIOS, RATES, AND PROPORTIONS

Section 3: Introduction to Proportions

A **proportion** is a statement that two ratios or rates are equivalent. For instance, the proportion $\frac{2}{3} = \frac{8}{12}$ states that the ratios $\frac{2}{3}$ and $\frac{8}{12}$ are equivalent. The proportion is read as "2 is to 3 as 8 is to 12."

In a proportion, the units in both numerators are the same and the units in both denominators are the same. In general, every proportion is of the form $\frac{a}{b} = \frac{c}{d}$, where a and c have the same units and b and d have the same units, and is read as "a is to b as c is to d."

EXAMPLE 1   Write a proportion.

(a) 6 inches is to 10 inches as 12 inches is to 20 inches
(b) $\frac{1}{2}$ is to 2 as $\frac{1}{16}$ is to $\frac{1}{4}$
(c) If it costs $78 for a 2-day car rental, then it costs $195 for a 5-day car rental.

SOLUTION

(a) In this problem, the common units, inches, divide out.

$$\frac{6 \text{ inches}}{10 \text{ inches}} = \frac{12 \text{ inches}}{20 \text{ inches}}$$

So, the proportion is $\frac{6}{10} = \frac{12}{20}$.

(b) $\dfrac{\frac{1}{2}}{2} = \dfrac{\frac{1}{16}}{\frac{1}{4}}$

(c) The proportion here is $78 is to 2 days as $195 is to 5 days.

$$\frac{\$78}{2 \text{ days}} = \frac{\$195}{5 \text{ days}}$$

PREALGEBRA REVIEW   CHAPTER 5 RATIOS, RATES, AND PROPOTIONS

One way to determine if a proportion is true is by checking that the two ratios in lowest terms are equal.

EXAMPLE 2   Is the proportion is true?

(a) $\dfrac{8}{14} = \dfrac{15}{21}$ (b) $\dfrac{27}{36} = \dfrac{9}{12}$

SOLUTION

(a) Write both fractions in lowest terms:

$$\dfrac{8}{14} = \dfrac{4}{7} \quad \text{and} \quad \dfrac{15}{21} = \dfrac{5}{7}$$

Since $\dfrac{4}{7}$ is not equal to $\dfrac{5}{7}$, the proportion is *not true*.

(b) In lowest terms, $\dfrac{27}{36} = \dfrac{3}{4}$ and $\dfrac{9}{12} = \dfrac{3}{4}$. Since both ratios are equivalent to $\dfrac{3}{4}$, the proportion is *true*.

Another way to determine if a proportion is true is by checking that the **cross products** are equal. A cross product is the product of the numerator of one fraction and the denominator of the other fraction. In the proportion $\dfrac{a}{b} = \dfrac{c}{d}$, the cross products are $a \cdot d$ and $c \cdot b$.

$$\dfrac{a}{b} \diagtimes \dfrac{c}{d} \quad \begin{matrix} c \cdot b \\ a \cdot d \end{matrix}$$

Therefore, if $c \cdot b = a \cdot d$, then the proportion is true.

CHAPTER 5   RATIOS, RATES, AND PROPORTIONS

EXAMPLE 3   Determine whether the proportion is true.

(a) $\dfrac{5}{4} = \dfrac{35}{28}$

(b) $\dfrac{\frac{1}{2}}{\frac{2}{3}} = \dfrac{6}{16}$

SOLUTION

(a) $\dfrac{5}{4} \diagdown\!\!\!\!\diagup \dfrac{35}{28}$ $\begin{array}{l} 35 \cdot 4 = 140 \\ \\ 5 \cdot 28 = 140 \end{array}$

Since the cross products are equal, the proportion is *true*.

(b) $\dfrac{\frac{1}{2}}{\frac{2}{3}} \diagdown\!\!\!\!\diagup \dfrac{6}{16}$ $\begin{array}{l} 6 \cdot \frac{2}{3} = \frac{12}{3} = 4 \\ \\ \frac{1}{2} \cdot 16 = \frac{16}{2} = 8 \end{array}$

Since the cross products are not equal, the proportion is *not true*.

PREALGEBRA REVIEW   CHAPTER 5   RATIOS, RATES, AND PROPOTIONS

Section 3: Exercises

Write a proportion. Do not simplify.

1. 3 is to 4 as 15 is to 20

2. 18 is to 32 as 9 is to 16

3. 8 liters is to 24 liters as 20 liters is to 60 liters

4. 21 days is to 42 days as 7 days is to 14 days

5. $4\frac{1}{2}$ inches is to 9 inches as $1\frac{1}{8}$ inches is to $2\frac{1}{4}$ inches

6. 6.25 meters is to 11 meters as 3.125 meters is to 5.5 meters

7. $165 is to 5 hours as $105 is to 3 hours

8. 208 miles is to 8 gallons as 286 miles is to 11 gallons

9. 32 feet is to 1 second as 288 is to 9 seconds

10. 25 yards is to 75 feet as 16 yards is to 48 feet

Determine whether the proportion is true.

11. $\dfrac{2}{5} = \dfrac{12}{30}$ 12. $\dfrac{18}{6} = \dfrac{9}{3}$ 13. $\dfrac{2}{3} = \dfrac{4}{5}$ 14. $\dfrac{21}{49} = \dfrac{3}{7}$

15. $\dfrac{55}{11} = \dfrac{25}{5}$ 16. $\dfrac{9}{72} = \dfrac{3}{24}$ 17. $\dfrac{63}{56} = \dfrac{7}{8}$ 18. $\dfrac{48}{12} = \dfrac{16}{8}$

19. $\dfrac{3.5}{3} = \dfrac{14}{12}$ 20. $\dfrac{15}{4.8} = \dfrac{10}{3.2}$ 21. $\dfrac{5}{\frac{3}{4}} = \dfrac{4}{\frac{3}{5}}$ 22. $\dfrac{\frac{1}{3}}{8} = \dfrac{\frac{1}{4}}{9}$

23. $\dfrac{50}{3\frac{1}{3}} = \dfrac{6}{\frac{2}{5}}$ 24. $\dfrac{4\frac{3}{8}}{\frac{5}{8}} = \dfrac{1\frac{1}{6}}{\frac{1}{6}}$ 25. $\dfrac{0.9}{1.8} = \dfrac{1.2}{3}$ 26. $\dfrac{0.4}{1.5} = \dfrac{2}{7.5}$

CHAPTER 5   RATIOS, RATES, AND PROPORTIONS   PREALGEBRA REVIEW

Section 4: Solving Proportions

Four numbers are used in a proportion. When one of the numbers is unknown, the other three can be used to find the unknown number so that the proportion is true.

Recall that if a proportion is true, then the cross products are equal. Therefore the cross products can be used to solve a proportion. For example, find the unknown number, represented by the variable x, that will make the following proportion true:

$$\frac{3}{8} = \frac{x}{56}$$

To solve the proportion, write an equivalent equation by finding the cross products and setting them equal. Then solve the resulting equation for x.

$$x \cdot 8 = 8x$$
$$3 \cdot 56 = 168$$

$8x = 168$	*Set the cross products equal.*
$\dfrac{8x}{8} = \dfrac{168}{8}$	*Divide each side by 8, the coefficient of x.*
$x = 21$	*Simplify.*

To check the answer, substitute the value for x in the original proportion and find the cross products.

$$\frac{3}{8} = \frac{21}{56} \qquad \textit{Substitute 21 for x.}$$

$$21 \cdot 8 = 168$$
$$3 \cdot 56 = 168$$

The cross products are equal, so 21 is the correct solution.

PREALGEBRA REVIEW   CHAPTER 5   RATIOS, RATES, AND PROPOTIONS

EXAMPLE 1   Solve and check.

(a) $\dfrac{12}{4} = \dfrac{9}{x}$ (b) $\dfrac{n}{7} = \dfrac{24}{28}$ (c) $\dfrac{4}{x} = \dfrac{8}{5}$

SOLUTION

(a)   SOLVE: $\dfrac{12}{4} \times \dfrac{9}{x}$ $\begin{array}{l} 9 \cdot 4 = 36 \\ 12 \cdot x = 12x \end{array}$ CHECK: $\dfrac{12}{4} \times \dfrac{9}{3}$ $\begin{array}{l} 9 \cdot 4 = 36 \\ 12 \cdot 3 = 36 \end{array}$

$36 = 12x$

$\dfrac{36}{12} = \dfrac{12x}{12}$

$3 = x$

(b)   SOLVE: $\dfrac{n}{7} \times \dfrac{24}{28}$ $\begin{array}{l} 24 \cdot 7 = 168 \\ n \cdot 28 = 28n \end{array}$ CHECK: $\dfrac{6}{7} \times \dfrac{24}{28}$ $\begin{array}{l} 24 \cdot 7 = 168 \\ 6 \cdot 28 = 168 \end{array}$

$168 = 28n$

$\dfrac{168}{28} = \dfrac{28n}{28}$

$6 = n$

(c)   SOLVE: $\dfrac{4}{x} \times \dfrac{8}{5}$ $\begin{array}{l} 8 \cdot x = 8x \\ 4 \cdot 5 = 20 \end{array}$ CHECK: $\dfrac{4}{\frac{5}{2}} \times \dfrac{8}{5}$ $\begin{array}{l} 8 \cdot \dfrac{5}{2} = \dfrac{40}{2} = 20 \\ 4 \cdot 5 = 20 \end{array}$

$8x = 20$

$\dfrac{8x}{8} = \dfrac{20}{8}$

$x = \dfrac{5}{2}$ or $x = 2\dfrac{1}{2}$

CHAPTER 5 RATIOS, RATES, AND PROPORTIONS

EXAMPLE 2 What value of t makes the following proportion true?

$$\frac{6}{4} = \frac{2\frac{1}{2}}{t}$$

SOLUTION

$2\frac{1}{2} \cdot 4 = \frac{5}{2} \cdot \frac{4}{1} = \frac{20}{2} = 10$

$6 \cdot t = 6t$

$10 = 6t$

$\dfrac{10}{6} = \dfrac{6t}{6}$

$\dfrac{5}{3} = t \quad \text{or} \quad 1\dfrac{2}{3} = t$

EXAMPLE 3 If 18 is to 45 as n is to 35, then what is n?

SOLUTION Start by setting up a proportion. Then solve the proportion.

$\dfrac{18}{45} = \dfrac{n}{35}$

$n \cdot 45 = 45n$

$18 \cdot 35 = 630$

$45n = 630$

$\dfrac{45n}{45} = \dfrac{630}{45}$

$n = 14$

PREALGEBRA REVIEWCHAPTER 5 RATIOS, RATES, AND PROPOTIONS

EXAMPLE 4 The ratio of two quantities X and Y is always $4:7$. Find the value of Y when X is 1.

SOLUTIONSet up a proportion and solve.

$$\frac{4}{7} = \frac{1}{Y}$$

$$\frac{4}{7} \times \frac{1}{Y} \quad \begin{array}{l} 1 \cdot 7 = 7 \\ 4 \cdot Y = 4Y \end{array}$$

$$7 = 4Y$$

$$\frac{7}{4} = \frac{4Y}{4}$$

$$\frac{7}{4} = Y \quad \text{or} \quad 1\frac{3}{4} = Y$$

CHAPTER 5   RATIOS, RATES, AND PROPORTIONS   PREALGEBRA REVIEW

Section 4: Exercises

Solve the proportion. Then check the answer.

1. $\dfrac{x}{4} = \dfrac{12}{16}$

2. $\dfrac{15}{x} = \dfrac{5}{6}$

3. $\dfrac{2}{7} = \dfrac{14}{y}$

4. $\dfrac{20}{12} = \dfrac{y}{3}$

5. $\dfrac{8}{t} = \dfrac{16}{8}$

6. $\dfrac{25}{15} = \dfrac{15}{t}$

7. $\dfrac{n}{48} = \dfrac{8}{32}$

8. $\dfrac{9}{36} = \dfrac{n}{72}$

9. $\dfrac{6}{x} = \dfrac{5}{9}$

10. $\dfrac{8}{3} = \dfrac{x}{2}$

11. $\dfrac{a}{7} = \dfrac{9}{10}$

12. $\dfrac{8}{12} = \dfrac{1}{a}$

13. $\dfrac{\frac{1}{4}}{\frac{1}{3}} = \dfrac{x}{20}$

14. $\dfrac{x}{\frac{2}{5}} = \dfrac{35}{\frac{1}{2}}$

15. $\dfrac{1\frac{2}{3}}{n} = \dfrac{1}{9}$

16. $\dfrac{0.4}{2} = \dfrac{4}{t}$

17. $\dfrac{2.4}{x} = \dfrac{1.2}{4}$

18. $\dfrac{2.25}{1.5} = \dfrac{x}{100}$

Choose the correct answer.

19. Which equation is equivalent to $\dfrac{24}{n} = \dfrac{18}{54}$?

    (A) $432 = 54n$ (B) $18n = 1296$ (C) $24n = 972$
    (D) $18n = 78$ (E) $18n = 2454$

20. Which equation is equivalent to $\dfrac{4.8}{3.2} = \dfrac{x}{0.8}$?

    (A) $3.2x = 5.6$ (B) $4.8x = 2.56$ (C) $0.8x = 15.36$
    (D) $3.2x = 38.4$ (E) $3.2x = 3.84$

21. Which proportion has 1.5 as its solution?

    (A) $\dfrac{6}{8} = \dfrac{1}{x}$ (B) $\dfrac{6}{9} = \dfrac{1}{x}$ (C) $\dfrac{14}{7} = \dfrac{1}{x}$ (D) $\dfrac{5}{4} = \dfrac{x}{1}$ (E) $\dfrac{10}{8} = \dfrac{x}{1}$

PREALGEBRA REVIEW   CHAPTER 5   RATIOS, RATES, AND PROPOTIONS

22. What is the value of b in the proportion $\dfrac{2.7}{b} = \dfrac{\frac{1}{4}}{\frac{2}{9}}$?

    (A) 2.4   **(B)** 1.5   **(C)** 1.2   **(D)** 0.6   **(E)** 0.15

23. If t is to 21 as 72 is to 84, what is t?

    (A) 24.5   **(B)** 18   **(C)** 288   **(D)** 12   **(E)** 15

24. If 0.5 is to 3 as 4 is to n, what is n?

    (A) 12   **(B)** 6   **(C)** 24   **(D)** $\frac{2}{3}$ **(E)** 0.375

25. The ratio of the quantities P and Q is always 5 to 2. If Q is 10, what is P?

    (A) 15   **(B)** 4   **(C)** 25   **(D)** 1   **(E)** 30

26. The ratio of the quantities X and Y is always the same. If $Y = 8$ when $X = 18$, find the value of Y when $X = 36$?

    (A) 144   **(B)** 12   **(C)** 16   **(D)** 4   **(E)** 81

27. Solve the proportion $\dfrac{x}{y} = \dfrac{z}{5}$ for x.

    (A) $\dfrac{5y}{z}$ **(B)** $\dfrac{5z}{y}$ **(C)** $5yz$ **(D)** $\dfrac{yz}{5}$ **(E)** $\dfrac{5}{yz}$

28. Solve the proportion $\dfrac{2}{b} = \dfrac{a}{3}$ for b.

    (A) $\dfrac{2a}{3}$ **(B)** $\dfrac{6}{a}$ **(C)** $\dfrac{3}{2a}$ **(D)** $\dfrac{3a}{2}$ **(E)** $\dfrac{a}{6}$

CHAPTER 5   RATIOS, RATES, AND PROPORTIONS

Section 5: Problem Solving with Proportions

Problem situations involving ratios or rates can be solved using proportions. When setting up these proportions, a variable can be used to represent the unknown quantity.

EXAMPLE 1   A photo printer can print 8 four by six photos in 2 minutes. How long will it take to print 24 photos?

SOLUTION   The time it takes to print 8 photos is given. To find the time it takes to print 24 photos, set up a proportion, letting x represent the unknown time.

$$\frac{\text{Number of photos}}{\text{Number of seconds}} = \frac{\text{Number of photos}}{\text{Number of seconds}} \rightarrow \frac{8}{2} = \frac{24}{x}$$

$$\frac{8}{2} = \frac{24}{x}$$
$$48 = 8x$$
$$\frac{48}{8} = \frac{8x}{8}$$
$$6 = x$$

It will take 6 minutes for the photo printer to print 24 photos.

EXAMPLE 2   Fran ran 3 miles in 24 minutes. At this rate, how many miles can she run in one hour?

SOLUTION   The number of miles Fran ran in 24 minutes is given. To find the number of miles she can run in one hour, or 60 minutes at the same rate, set up a proportion, letting x represent the unknown number of miles.

$$\frac{\text{Number of miles}}{\text{Number of minutes}} = \frac{\text{Number of miles}}{\text{Number of minutes}} \rightarrow \frac{3}{24} = \frac{x}{60}$$

$$\frac{3}{24} = \frac{x}{60}$$
$$24x = 180$$
$$\frac{24x}{24} = \frac{180}{24}$$
$$x = 7.5$$

At that rate, she will run 7.5 miles in one hour.

PREALGEBRA REVIEW CHAPTER 5 RATIOS, RATES, AND PROPOTIONS

EXAMPLE 3 Mac earns $167.75 in 11 hours. How much does he earn in 40 hours?

SOLUTION The amount he earns in 11 hours is given. To find the amount he earns in 40 hours, set up a proportion, letting x represent the unknown amount he earns.

$$\frac{\text{Amount earned}}{\text{Number of hours}} = \frac{\text{Amount earned}}{\text{Number of hours}} \rightarrow \frac{\$167.75}{11} = \frac{\$x}{40}$$

$$\frac{167.75}{11} = \frac{x}{40}$$
$$11x = 6710$$
$$\frac{11x}{11} = \frac{6710}{11}$$
$$x = 610$$

Mac earns $610 in 40 hours.

EXAMPLE 4 An elementary school has a student to teacher ratio of 16 to 1. If 208 students attend the school, how many teachers teach at the school?

SOLUTION The student to teacher ratio of 16 to 1 or $\frac{16}{1}$ is given. To find the number of teachers teaching at the school, set up proportion, letting x represent the unknown number of teachers.

$$\frac{16}{1} = \frac{\text{Total number of students}}{\text{Total number of teachers}} \rightarrow \frac{16}{1} = \frac{208}{x}$$

$$\frac{16}{1} = \frac{208}{x}$$
$$208 = 16x$$
$$\frac{208}{16} = \frac{16x}{16}$$
$$13 = x$$

So 13 teachers teach at the school.

CHAPTER 5   RATIOS, RATES, AND PROPORTIONS   PREALGEBRA REVIEW

EXAMPLE 5   A gallon of paint covers 400 square feet of wall. How much paint is needed to cover 500 square feet of wall with two coats of paint?

SOLUTION   The number of square feet one gallon of paint covers is given. To find the number of gallons of paint needed to cover 500 square feet with two coats can be found by setting up a proportion. Since two coats of paint are to be applied, the total number of square feet the paint must cover is 2×500 square feet, or 1000 square feet.

$$\frac{\text{Number of square feet}}{\text{Number of gallons}} = \frac{\text{Number of square feet}}{\text{Number of gallons}} \rightarrow \frac{400}{1} = \frac{1000}{x}$$

$$\frac{400}{1} = \frac{1000}{x}$$
$$1000 = 400x$$
$$\frac{1000}{400} = \frac{400x}{400}$$
$$2.5 = x$$

So 2.5 gallons of paint are needed to cover 500 square feet of wall with two coats.

EXAMPLE 6   A solution contains 2 ounces of alcohol and 7.5 ounces of water. If another solution containing 3 ounces of water is to have the same ratio of ounces of alcohol to ounces of water, how much alcohol should it contain?

SOLUTION   The amount of alcohol in 7.5 ounces of water is given. To find the amount of alcohol needed in 3 ounces of water, set up a proportion.

$$\frac{\text{Amount of alcohol}}{\text{Amount of water}} = \frac{\text{Amount of alcohol}}{\text{Amount of water}} \rightarrow \frac{2 \text{ ounces}}{7.5 \text{ ounces}} = \frac{x \text{ ounces}}{3 \text{ ounces}}$$

$$\frac{2}{7.5} = \frac{x}{3}$$
$$7.5x = 6$$
$$\frac{7.5x}{7.5} = \frac{6}{7.5}$$
$$x = 0.8$$

The solution should contain 0.8 ounce of alcohol.

PREALGEBRA REVIEW   CHAPTER 5   RATIOS, RATES, AND PROPOTIONS

Section 5: Exercises

Choose the correct answer.

1. If 1 inch on a map represents 5 miles, then how many miles apart are two cities if the distance between them on a map is 3 inches?

    (A) $\dfrac{3}{5}$ (B) 15   (C) 3   (D) $\dfrac{5}{3}$ (E) 8

2. If 2 tablespoons of coffee yields 6 fluid ounces, how many tablespoons of coffee are needed to yield 48 fluid ounces?

    (A) 8   (B) 24   (C) 16   (D) 4   (E) 6

3. In a statewide survey, 8 out of 10 people said they would recommend their doctor to another person. If 420 people were surveyed, how many said they would not recommend their doctor?

    (A) 336   (B) 525   (C) 42   (D) 84   (E) 378

4. At a local community college, 3 out of 8 students are enrolled in allied health programs. If 816 students are enrolled in allied health programs, how many students attend the college?

    (A) 2176   (B) 306   (C) 1360   (D) 2448   (E) 6528

5. A basketball player makes 15 baskets in 21 attempts. At this rate, how many baskets will he miss in 70 attempts?

    (A) 30   (B) 20   (C) 50   (D) 294   (E) 98

6. A baseball player gets 6 hits for every 25 at bats. At this rate, how many hits will he get in 150 at bats?

    (A) 114   (B) 48   (C) 25   (D) 31   (E) 36

7. A new car can travel 24 miles for each gallon of gas. How many gallons of gas will the car need to travel 288 miles?

    (A) 14   (B) 18   (C) 12   (D) 15   (E) 10

CHAPTER 5 RATIOS, RATES, AND PROPORTIONS   PREALGEBRA REVIEW

8. During a sale, a store offers customers $5 off for every $50 they spend. If a customer spent $200, how much did the store charge after the discount was applied?

   (A) $20   (B) $120   (C) $150   (D) $180   (E) $80

9. A prescription calls for 50 milligrams of medication to be administered every 4 hours in liquid form. If there are 8 milligrams of medication per 12 milliliters of solution, how many milliliters should the patient receive every 4 hours?

   (A) 200   (B) 75   (C) 48   (D) 100   (E) 60

10. A quality control inspector finds that for every 500 parts manufactured, 2 are defective. At this rate, how many of the 2500 parts produced in one day will not be defective?

    (A) 15   (B) 10   (C) 2490   (D) 1000   (E) 500

11. For every half hour of jogging, a 150 pound person can burn 238 calories. At this rate, how many calories will a 150 pound person burn if he jogs for 45 minutes?

    (A) 100   (B) 714   (C) 324   (D) 450   (E) 357

12. An instructor must grade 40 math exams. If she can grade $\frac{1}{5}$ of the exams in 45 minutes, how long will it take her to grade all 40 exams?

    (A) 360 min   (B) 90 min   (C) 225 min   (D) 135 min   (E) 180 min

13. Aida invested her money into a low-risk fund and a high-risk fund in a ratio of 3 to 2, respectively. If she invested $850 in the high-risk fund, how much more did she invest in the low-risk fund?

    (A) $425.00   (B) $567.67   (C) $1275.00   (D) $283.33   (E) $550.00

14. An electronics company manufactures flat screen plasma televisions. The length to the height of the screen of all the televisions are in a ratio of 16 to 9. What is the length of the screen if the height is 16.2 inches?

    (A) 9.1125 in   (B) 28.8 in   (C) 8.9 in   (D) 32.4 in   (E) 24.8 in

15. A two-stroke cycle engine requires a gasoline oil mixture in a ratio of 40 to 1. How many ounces of oil should be added to 192 ounces of gasoline to obtain the correct mixture?

    (A) 6.5   (B) 152   (C) 5.2   (D) 7680   (E) 4.8

PREALGEBRA REVIEW   CHAPTER 5   RATIOS, RATES, AND PROPOTIONS

16. A commuter spends $23 for 10 rides on a commuter train. How much will it cost for 25 rides?

   (A) $48   (B) $57.50   (C) $92   (D) $108   (E) $35

17. If 7 amusement park admissions cost $315, how much will 3 admissions cost?

   (A) $15   (B) $105   (C) $135   (D) $45   (E) $185

18. Martin earns one sick day for every fifteen days he works. How many sick days would he have after working 180 days?

   (A) 30   (B) 12   (C) 18   (D) 15   (E) 20

19. A grocery store sells canned vegetables at 3 for $2. At this rate, how much will 8 cans cost? (Round to the nearest cent.)

   (A) $5.00   (B) $0.67   (C) $5.33   (D) $12.00   (E) $5.67

20. A produce store sells grapes for $1.09 per pound. To the nearest cent, how much will it cost for $\frac{3}{4}$ pound of grapes?

   (A) $0.82   (B) $0.80   (C) $0.89   (D) $0.92   (E) $0.81

21. An athlete can run 2.4 kilometers in 22.8 minutes. At this rate, how many kilometers can she run in 38 minutes?

   (A) 9.5   (B) 1.44   (C) 4   (D) 3.8   (E) 3.2

22. The EPA highway mileage estimate for a new sedan is listed as 32 miles per gallon. If the sedan has a gas tank with a capacity of 18.2 gallons, how many miles can it be driven on a full tank of gas?

   (A) 328.4   (B) 50.2   (C) 320   (D) 582.4   (E) 486.4

23. In a math class, 5 out of 8 students are men. If there are 48 students in the class, how many more men than women are there?

   (A) 30   (B) 18   (C) 9   (D) 16   (E) 12

CHAPTER 5   RATIOS, RATES, AND PROPORTIONS   PREALGEBRA REVIEW

24. In a recent survey of 234 high school students, 7 out of 9 said they planned on going to college after graduation. How many more students are planning to go to college after graduation than those not planning to go?

   (A) 52   (B) 130   (C) 182   (D) 78   (E) 126

25. The length and width of a rectangle are in a ratio of 5 to 3 respectively. Find the width if the length is 4.5 inches.

   (A) 2.7 in   (B) 7.5 in   (C) 0.9 in   (D) 1.5 in   (E) 2.5 in

26. In a certain town, the residential property tax rate is $9.92 per $1000 of assessed value of a home. At this rate, how much more property tax is paid on a home with an assessed value of $234,000 than on a home with an assessed value of $156,000?

   (A) $2321.28   (B) $234.00   (C) $773.76   (D) $826.24   (E) $547.52

27. The nutrition information on a box of cereal says that a $\frac{3}{4}$ cup serving contains 120 calories and 22 grams of carbohydrates. At this rate, how many calories and grams of carbohydrates are in a $\frac{1}{2}$ cup serving?

   (A) $45; 8\frac{1}{4}$ (B) 60; 11   (C) $90; 16\frac{1}{2}$ (D) $80; 14\frac{2}{3}$ (E) 60; 16

28. A consumer survey found that 5 out of 6 people drink coffee. Of the people who drink coffee, 1 out of 12 drink decaffeinated coffee only. How many out of 90,000 people would be expected to only drink decaffeinated coffee?

   (A) 7500   (B) 6250   (C) 75,000   (D) 1250   (E) 250

Chapter 6 Percents

Section 1: Introduction to Percent

A **percent** is a fraction or ratio with a denominator of 100. Percent means "per 100" or "parts out of 100." The symbol % is used to denote percent. For example, if a student earns 85 points out of 100 possible points on an exam, then the student's score is $\frac{85}{100}$ or 85% (read as "eighty-five percent").

EXAMPLE 1 Write in percent notation.

(a) 47 out of 100 employees are women
(b) The sales tax rate is $5 for every $100 spent
(c) A company survey of 100 employees found that all 100 felt overworked

SOLUTION

(a) $\frac{47}{100} = 47\%$

(b) $\frac{5}{100} = 5\%$

(c) $\frac{100}{100} = 100\%$

Note that 100% of something indicates the whole. In Example 1(c), 100% indicates the whole group of employees surveyed.

EXAMPLE 2 Write in fraction notation. Do not simplify.

(a) A student got 89% of the questions correct on an exam
(b) The insect repellant contains 10% DEET

SOLUTION

(a) $\frac{89}{100}$

(b) $\frac{10}{100}$

EXAMPLE 3 Of the students enrolled in a math class, 48% are men. What percent of the students enrolled in the math class are women?

SOLUTION The percent of men enrolled is given. Remember that 100% represents the whole class. To find the percent of students enrolled who are women, subtract.

$$\begin{array}{c}\text{Percent representing}\\\text{whole class}\end{array} - \begin{array}{c}\text{Percent of men}\\\text{enrolled}\end{array} \rightarrow 100\% - 48\%$$

$$100 - 48 = 52$$

So 52% of the students enrolled in the class are women.

Consider the two percents 31% and 67%. To determine which percent is greater, the percents can be written in fraction form and then compared.

$$31\% = \frac{31}{100} \quad \text{and} \quad 67\% = \frac{67}{100}$$

Since $31 < 67$, $\frac{31}{100} < \frac{67}{100}$. So 67% is greater than 31%.

In general, to compare two percents, drop the percent symbol and compare the numbers.

EXAMPLE 4 Joe's Visa card has an annual percentage rate (APR) of 21%, whereas his MasterCard has an APR of 18%. Which credit card has a higher APR?

SOLUTION Since $21 > 18$, $21\% > 18\%$. So his Visa card has a higher APR.

PREALGEBRA REVIEW   CHAPTER 6   PERCENTS

Section 1: Exercises

Write in percent notation.

1. 68 milliliters out of a 100 milliliter solution is alcohol

2. 1 out of every 100 parts manufactured were defective

3. A student received 83 points out of 100 points on an exam

4. A store gives a $15 discount for every $100 spent

5. 91 miles of a 100 mile trip were driven on the freeway

6. 43 out of 100 babies born at a hospital were boys

Write in fraction notation. Do not simplify.

7. 23% of the students in a school were absent today

8. 61% of Gina's monthly earnings are used to pay her rent and bills

9. A restaurant includes an 18% gratuity for parties of 8 or more

10. 57% of the cars sold at a local dealership last year were sedans

11. An employee received a 6% pay increase this year

12. 89% of a utility company's customers lost power during a storm

Solve.

13. A company determines that 26% of its revenue is generated from online sales. What percent of the company's revenue is not generated from online sales?

14. Of the degrees awarded at commencement, 68% were Associate's degrees and the rest were Bachelor's degrees. What percent of the degrees awarded were Bachelor's degrees?

CHAPTER 6   PERCENTS

15. A student answered 88% of the questions on an exam correctly. What percent of the questions did he answer incorrectly?

16. Of the athletes who started a marathon, 19% did not finish. What percent of the athletes finished the marathon?

17. Of two brands of juice, Brand A contains 100% fruit juice and Brand B contains 30% fruit juice. Which brand contains less fruit juice?

18. During a holiday sale, Computer City advertises 25% off all laptop computers, whereas PC Mart advertises 15% off all laptop computers. Which store offers a higher discount?

19. Anytown Bank offers 2.9% interest on a savings account, whereas Midstate Bank offers 3% interest on a savings account. Which bank offers a higher interest rate?

20. Topical ointment X contains 0.5% hydrocortisone and topical ointment Y contains 1% hydrocortisone. Which topical ointment contains more hydrocortisone?

PREALGEBRA REVIEW
CHAPTER 6 PERCENTS

Section 2: Percents and Decimals

Converting Percents to Decimals

Recall from Chapter 3 that the fraction $\frac{13}{100}$ is read as "thirteen hundredths" and can be written in decimal form as 0.13. This fact together with the meaning of percent can be used to convert a percent to a decimal. For instance, 35% means 35 out of 100 and can be written as $\frac{35}{100}$. Since the fraction bar indicates division, $\frac{35}{100}$ can be written as 35 ÷ 100. The result of this division is the decimal 0.35.

Procedure for Converting a Percent to a Decimal:
Drop the % symbol and divide the number by 100.

EXAMPLE 1 Convert the decimal to a percent.

  (a) 56%   (b) 2%

SOLUTION

  (a) 56% = 56 ÷ 100 = 0.56
  (b) 2% = 2 ÷ 100 = 0.02

Recall from Chapter 3 that dividing by a power of 10 moves the decimal point in the original number to the *left* the same number of places as the number of zeros in the power of 10. So a "shortcut" for converting a percent to a decimal is to drop the percent symbol and move the decimal point two places to the left.

EXAMPLE 2 Write the percent as a decimal.

  (a) 80%   (b) 145%   (c) 25.9%   (d) 3.1%   (e) 0.25%

SOLUTION

  (a) 80% = 0.80, or 0.8
  (b) 145% = 1.45
  (c) 25.9% = 0.259
  (d) 3.1% = 0.031
  (e) 0.25% = 0.0025

CHAPTER 6 PERCENTS

Converting Decimals to Percents

To convert a decimal to a percent, the process that was used to convert a percent to a decimal is reversed.

> **Procedure for Converting a Decimal to a Percent:**
> Multiply the decimal by 100 and attach a percent symbol to the right of the number.

EXAMPLE 3   Convert the decimal to a percent.

(a) 0.4   (b) 0.08

SOLUTION

(a) $0.4 = 0.4 \times 100 = 40\%$
(b) $0.08 = 0.08 \times 100 = 8\%$

Recall from Chapter 3 that multiplying by a power of 10 moves the decimal point in the original number to the *right* the same number of places as the number of zeros in the power of 10. So a "shortcut" for converting a decimal to a percent is to move the decimal point two places to the right and then attach the % symbol to the right of the number.

EXAMPLE 4   Write the decimal as a percent.

(a) 0.67   (b) 2.5   (c) 0.888   (d) 0.076   (e) 0.001

SOLUTION

(a) $0.67 = 67\%$
(b) $2.5 = 250\%$
(c) $0.888 = 88.8\%$
(d) $0.076 = 7.6\%$
(e) $0.001 = 0.1\%$

Notice that decimal numbers larger than 1 result in percents that are larger than 100%.

Section 2: Exercises

Convert the percent to a decimal.

1. 72%   2. 13%   3. 90%   4. 10%

5. 5%   6. 3%   7. 150%   8. 133%

9. 4.8%   10. 9.67% 11. 0.2% 12. 0.04%

13. 18.35% 14. 68.1% 15. 0.016% 16. 0.207%

Convert the decimal to a percent.

17. 0.48   18. 0.22   19. 0.7   20. 0.6

21. 0.09   22. 0.01   23. 0.375 24. 0.903

25. 1.66   26. 2.1   27. 0.033 28. 0.009

29. 0.5515 30. 0.0718 31. 0.0046 32. 0.0008

Choose the correct answer.

33. What is 30.8% in decimal form?

    (A) 3.08   (B) 30.8   (C) 0.308   (D) 0.038   (E) 3088

34. What is 0.0101 written as a percent?

    (A) 1.01%   (B) 10.1%   (C) 101%   (D) 0.101%   (E) 0.000101%

35. A bank offers an annual interest rate of 2.3% for its savings accounts. Write this percent as a decimal.

    (A) 2.3   (B) 0.23   (C) 230   (D) 0.023   (E) 20.3

36. A college bookstore marks up the price of a math textbook. The selling price is 135% of the original price of the book. What is this percent expressed as a decimal?

    (A) 13.5   (B) 1.35   (C) 0.135   (D) 135   (E) 0.0135

Section 3: Percents and Fractions

Converting Percents to Fractions

The meaning of percent is used to convert a percent to a fraction. For instance, 35% means 35 out of 100 and can be written as $\frac{35}{100}$.

> **Procedure for Converting a Percent to a Fraction:**
> Drop the % symbol and place the number over 100. Then simplify the fraction.

EXAMPLE 1 Convert the percent to a fraction.

(a) 25%   (b) 60%   (c) 56%   (d) 140%

SOLUTION

(a) $25\% = \dfrac{25}{100} = \dfrac{\cancel{25}\cdot 1}{\cancel{25}\cdot 4} = \dfrac{1}{4}$

(b) $60\% = \dfrac{60}{100} = \dfrac{\cancel{20}\cdot 3}{\cancel{20}\cdot 5} = \dfrac{3}{5}$

(c) $56\% = \dfrac{56}{100} = \dfrac{\cancel{4}\cdot 14}{\cancel{4}\cdot 25} = \dfrac{14}{25}$

(d) $140\% = \dfrac{140}{100} = \dfrac{\cancel{20}\cdot 7}{\cancel{20}\cdot 5} = \dfrac{7}{5}$

PREALGEBRA REVIEW CHAPTER 6 PERCENTS

EXAMPLE 2   Convert 85.5% to a fraction.

SOLUTION

$$85.5\% = \frac{85.5}{100}$$
$$= \frac{85.5 \cdot 10}{100 \cdot 10} \quad \textit{Multiply the numerator and denominator by 10 to eliminate the decimal point in the numerator.}$$
$$= \frac{855}{1000} = \frac{\cancel{5} \cdot 171}{\cancel{5} \cdot 200} = \frac{171}{200}$$

An alternative way to convert percents to fractions is to first rewrite the percent in decimal form and then convert the decimal to a fraction. For instance,

$$85.5\% = 0.855 = \frac{855}{1000} = \frac{\cancel{5} \cdot 171}{\cancel{5} \cdot 200} = \frac{171}{200}$$

Note that this is the same result obtained in Example 2.

EXAMPLE 3   Write 2.25% as a fraction.

SOLUTION   Using the alternative method for converting fractions gives:

$$2.25\% = 0.0225 = \frac{225}{10{,}000} = \frac{\cancel{25} \cdot 9}{\cancel{25} \cdot 400} = \frac{9}{400}$$

It may be helpful to review complex fractions in Chapter 2 before reading through Example 4.

EXAMPLE 4   Write $33\frac{1}{3}\%$ in fraction form.

SOLUTION

$$33\frac{1}{3}\% = \frac{33\frac{1}{3}}{100}$$
$$= 33\frac{1}{3} \div 100 = \frac{100}{3} \div 100 = \frac{100}{3} \cdot \frac{1}{100} = \frac{\cancel{100} \cdot 1}{3 \cdot \cancel{100}} = \frac{1}{3}$$

Converting Fractions to Percents

Recall from Chapter 3 that to convert a fraction to a decimal, write the fraction as a division problem and then divide. This procedure will be used to convert fractions to percents.

> **Procedure for Converting a Fraction to a Percent:**
> Write the fraction in decimal form. Then convert the decimal to a percent.

EXAMPLE 5   Write $\dfrac{4}{5}$ as a percent.

SOLUTION   First write the fraction in decimal form by dividing the numerator by the denominator.

$$\frac{4}{5} = 5\overline{)4.0} \quad \begin{array}{r} 0.8 \\ \underline{4\,0} \\ 0 \end{array}$$

Then convert 0.8 to a percent: $0.8 = 80\%$

EXAMPLE 6   Convert the fraction to a percent.

(a) $\dfrac{1}{8}$ (b) $\dfrac{12}{25}$

SOLUTION

(a) $\dfrac{1}{8} = 8\overline{)1.000} \quad \begin{array}{r} 0.125 \\ \underline{8} \\ 20 \\ \underline{16} \\ 40 \\ \underline{40} \\ 0 \end{array}$

(b) $\dfrac{12}{25} = 25\overline{)12.00} \quad \begin{array}{r} 0.48 \\ \underline{10\,0} \\ 2\,00 \\ \underline{2\,00} \\ 0 \end{array}$

$0.125 = 12.5\%$ $0.48 = 48\%$

PREALGEBRA REVIEW   CHAPTER 6   PERCENTS

Sometimes a repeating decimal occurs when converting a fraction to a percent. When this happens, carry out the division so that there are enough decimal places to allow for rounding to an indicated decimal place. For instance, to round a percent to the nearest tenth, the division should be carried out to four decimal places as shown in Example 7.

EXAMPLE 7   Convert $\frac{2}{9}$ to a percent rounded to the nearest tenth.

SOLUTION

$$\frac{2}{9} = 9\overline{)2.0000}^{\,0.2222}$$

$$\begin{array}{r} \underline{1\,8} \\ 20 \\ \underline{18} \\ 20 \\ \underline{18} \\ 20 \\ \underline{18} \\ 2 \end{array}$$

$$\frac{2}{9} = 0.2222\ldots \text{ and } 0.2222 = 22.2\%$$

EXAMPLE 8   Convert $1\frac{1}{5}$ to a percent.

SOLUTION

$$1\frac{1}{5} = \frac{6}{5} \quad \rightarrow \quad \frac{6}{5} = 5\overline{)6.0}^{\,1.2}$$

$$\begin{array}{r} \underline{5} \\ 10 \\ \underline{10} \\ 0 \end{array}$$

$$1.2 = 120\%$$

CHAPTER 6   PERCENTS

Section 3: Exercises

Convert the percent to a fraction.

1. 36%   2. 95%   3. 20%   4. 30%

5. 225%   6. 110%   7. 6%   8. 5%

9. 12.8%   10. 52.2%   11. 87.5%   12. 0.25%

13. $44\frac{4}{9}\%$ 14. $66\frac{2}{3}\%$ 15. $50\frac{2}{5}\%$ 16. $\frac{1}{2}\%$

Convert the fraction or mixed number to a percent.

17. $\frac{1}{2}$ 18. $\frac{3}{4}$ 19. $\frac{5}{8}$ 20. $\frac{13}{20}$

21. $\frac{9}{16}$ 22. $\frac{33}{50}$ 23. $\frac{7}{10}$ 24. $\frac{24}{25}$

25. $\frac{3}{2}$ 26. $\frac{11}{8}$ 27. $2\frac{1}{10}$ 28. $1\frac{4}{5}$

Write the fraction as a percent rounded to the nearest hundredth.

29. $\frac{1}{6}$ 30. $\frac{1}{12}$

31. $\frac{8}{9}$ 32. $\frac{13}{15}$

33. $\frac{4}{7}$ 34. $\frac{9}{11}$

35. $\frac{28}{45}$ 36. $\frac{7}{18}$

PREALGEBRA REVIEW

CHAPTER 6   PERCENTS

Section 4: Percent Equations

As with fractions and decimals, percents are used to describe parts of a whole. The whole is called the **base**, and the part being compared to the base is called the **amount.**

In general, the percent equation is

$$\text{Percent} \times \text{Base} = \text{Amount}$$

Note that three numbers are used in the percent equation: percent, base, and amount. If one of the numbers is unknown, the other two can be used to find the unknown number.

In translating percent problems to an equation, replace the **what** with a variable, replace **of** with a multiplication symbol, and replace **is** with an equal sign.

Depending on which of the three numbers is missing, one of the following three equations will be used to solve a percent problem. Note that $p\%$ represents percent, a represents the amount, and b represents the base in the equations.

1. Finding the amount, a: "What number is $p\%$ of b?" or "Find $p\%$ of b."
2. Finding the base, b: "a is $p\%$ of what number?"
3. Finding the percent, p: "What percent of b is a?" or "b is what percent of a?"

Note that to perform calculations with percents, change the percent to a decimal before solving.

EXAMPLE 1   Translate and solve: 96 is 25% of what number?

SOLUTION   Use the equation for finding the base.

Translate: 96 is 25% of what number
  ↓   ↓   ↓   ↓   ↓
  96 = 25% × b

Solve: To solve, change 25% to 0.25 and divide both sides by 0.25.

$$96 = 0.25 \times b$$
$$\frac{96}{0.25} = \frac{0.25 \times b}{0.25}$$
$$384 = b$$

CHAPTER 6   PERCENTS   PREALGEBRA REVIEW

EXAMPLE 2   Find 42% of 75?

SOLUTION   Use the equation for finding the amount.

Translate: $a = 42\% \cdot 75$

Solve: To solve, change 42% to 0.42 and multiply.

$a = 0.42 \cdot 75$
$a = 31.5$

EXAMPLE 3   What percent of 120 is 102?

SOLUTION   Use the equation for finding the percent.

Translate: $p\% \cdot 120 = 102$

Solve: To solve, divide both sides by 120.

$p\% \cdot 120 = 102$
$\dfrac{p\% \cdot 120}{120} = \dfrac{102}{120}$
$p\% = 0.85$
$p = 85\%$ *Convert the decimal to a percent.*

EXAMPLE 4   282.9 is what percent of 246?

SOLUTION   Use the equation for finding the percent.

Translate: $282.9 = p\% \cdot 246$

Solve: To solve, divide both sides by 246.

$282.9 = p\% \cdot 246$
$\dfrac{282.9}{246} = \dfrac{p\% \cdot 246}{246}$
$1.15 = p\%$
$115\% = p$

PREALGEBRA REVIEW   CHAPTER 6   PERCENTS

Section 4: Exercises

Solve.

1. What percent of 50 is 28?

2. What number is 36% of 150?

3. Find 75% of 24.

4. 15 is 20% of what number?

5. 30.6 is what percent of 45?

6. 198 is 99% of what number?

7. 9.6 is what percent of 6.4?

8. Find 130% of 78.

9. 270 is what percent of 90?

10. What percent of 10 is 0.1?

11. 6 is 12.5% of what number?

12. What number is 84.9% of 400?

Choose the correct answer.

13. What number is 0.5% of 18?

    (A) 9   (B) 0.09   (C) 0.9   (D) 90   (E) 0.009

14. 10.24 is what percent of 256?

    (A) 4%   (B) 25%   (C) 2500%   (D) 40%   (E) 250%

CHAPTER 6   PERCENTS

15. $1\frac{3}{5}$ is 48% of what number?

(A) $3\frac{3}{4}$ (B) 3   (C) $\frac{96}{125}$ (D) $3\frac{3}{10}$ (E) $3\frac{1}{3}$

16. Find $3\frac{1}{4}\%$ of 128.

(A) 41.6   (B) 4.16   (C) 4.8   (D) 416   (E) 4.48

17. What percent of 30 is 20?

(A) 150%   (B) $33\frac{1}{3}\%$ (C) $66\frac{2}{3}\%$ (D) 15%   (E) 66%

18. 0.6 is what percent of $\frac{3}{4}$?

(A) 20%   (B) 125%   (C) 12.5%   (D) 80%   (E) 8%

19. If 50% of a number is 22, then what is 25% of the number?

(A) 11   (B) 5.5   (C) 36   (D) 2.75   (E) 44

20. If 6% of a number is 94.8, then 1422 is what percent of the number?

(A) 900%   (B) 11%   (C) 80%   (D) 90%   (E) 111%

PREALGEBRA REVIEW

CHAPTER 6 PERCENTS

Section 5: Problem Solving with Percents

Many real-life situations involve percents. When solving applied problems involving percents, the key is to determine which quantity, the amount, the base, or the percent, is to be found.

EXAMPLE 1 In a building that has 50 condominium units, 84% are owner occupied. How many of the units are owner occupied?

SOLUTION The number of condominium units (the base) and the percent of the units that are owner occupied are given. To find the number of units, use the equation for finding the amount.

$$\text{Number of owner occupied units} = 84\% \cdot \text{Total number of units} \quad \rightarrow \quad a = 84\% \cdot 50$$

$$a = 0.84 \cdot 50 = 42$$

So, 42 of the units are owner occupied.

EXAMPLE 2 A music CD has 15 songs. Twelve songs are under four minutes long. What percent of the songs on the CD are under four minutes long?

SOLUTION The total number of songs on the CD (the base) and the number of songs that are under four minutes long (the amount) are given. To solve, use the equation for finding the percent.

$$\text{Number of songs under four minutes} = p\% \cdot \text{Total number of songs on CD} \quad \rightarrow \quad 12 = p\% \cdot 15$$

$$12 = p\% \cdot 15$$
$$\frac{12}{15} = \frac{p\% \cdot 15}{15}$$
$$0.8 = p\%$$
$$80\% = p$$

So, 80% of the songs on the CD are under four minutes long.

CHAPTER 6 PERCENTS   PREALGEBRA REVIEW

EXAMPLE 3   A couple pays $57,800 as a down payment, which represents 20% of the purchase price of a home. What is the purchase price of the home?

SOLUTION   The amount of the down payment (the amount) and the percent are given. To find the purchase price of the home, use the equation for finding the base.

Down payment = 20% · Purchase price   → $57,800 = 20% · b

$$57,800 = 0.20 \cdot b$$
$$\frac{57,800}{0.20} = \frac{0.20 \cdot b}{0.20}$$
$$289,000 = b$$

The purchase price of the home was $289,000.

EXAMPLE 4   A car dealership has compact cars, sedans, and SUV's on its lot. There are a total of 240 vehicles on the lot. One-fourth of the vehicles are compact cars. Of the remaining vehicles, 40% are SUV's. What percent of the vehicles on the lot are sedans?

SOLUTION   The total number of vehicles is given. To find the percent of the vehicles on the lot that are sedans, first find the number of vehicles that are compact cars and then find the number of vehicles that are SUV's.

The problem states that $\frac{1}{4}$ of the vehicles are compact cars. To determine the number of vehicles that are compact cars, multiply.

$$\frac{1}{4} \cdot \text{total number of vehicles} \quad \rightarrow \quad \frac{1}{4} \cdot 240: \quad \frac{1}{4} \cdot 240 = 60$$

There are 60 compact cars. That means that 240 − 60, or 180 of the vehicles are either sedans or SUV's. The problem states that 40% of the remaining 180 vehicles are SUV's. To find the number of SUV's, use the equation for finding the amount.

Number of SUV's = 40% · 180   → $a = 40\% \cdot 180$

$$a = 0.4 \cdot 180 = 72$$

There are 72 SUV's. That means that of the 180 vehicles that remain, 180 − 72, or 108 are sedans.

PREALGEBRA REVIEW   CHAPTER 6   PERCENTS

(Example 4 continued)

To find the percent of the vehicles on the lot that are sedans, use the equation for finding the percent.

Number of sedans = $p\%$ · Total number of vehicles   → $108 = p\% \cdot 240$

$$108 = p\% \cdot 240$$
$$\frac{108}{240} = \frac{p\% \cdot 240}{240}$$
$$0.45 = p\%$$
$$45\% = p$$

So, 45% of the vehicles on the lot are sedans.

EXAMPLE 5   On an exam a student answered 95% of the 20 multiple choice questions correctly and 50% of the 16 short answer questions correctly. What percent of the 36 exam questions did the student answer correctly?

SOLUTION   The percents of the multiple choice questions and the short answer questions the student answered correctly and the number of each type of question on the exam are given. To find the percent of the exam questions the student answered correctly, first find the number of each type of question were answered correctly. To do this, use the equation for finding the amount.

Multiple choice: $a = 0.95 \cdot 20 = 19$ Short answer: $a = 0.5 \cdot 16 = 8$

The student answered 19 of the multiple choice and 8 of the short answer questions correctly. So the student answered 19 + 8, or 27 of the exam questions correctly. Now use the equation for finding the percent.

Number of questions answered correctly = $p\%$ · Number of questions on exam   → $27 = p\% \cdot 36$

$$27 = p\% \cdot 36$$
$$\frac{27}{36} = \frac{p\% \cdot 36}{36}$$
$$0.75 = p\%$$
$$75\% = p$$

The student answered 75% of the exam questions correctly.

Percent Increase and Percent Decrease

Many applied problems dealing with increases or decreases in prices, populations, earnings, and other values involve finding the percent of change. A percent increase tells how much a value has increased over the original value and a percent decrease tells how much a value has decreased over the original value.

To find the percent increase (or decrease), first subtract to find the amount of increase (or decrease):

Amount of increase: New value – original value
Amount of decrease: Original value – new value

Then, use the equation for finding the percent: a is what percent of b? Note that in this situation, a represents the amount of increase (or decrease), and b represents the original value.

EXAMPLE 6 Mae's hourly wage was raised from $12.80 to $13.44. What was the percent increase in her hourly wage?

SOLUTION Mae's original hourly wage and her new wage are given. To find the percent increase, first find the amount of increase by subtracting.

New wage – Original wage   → $13.44 – $12.80

13.44 – 12.80 = 0.64

The amount of increase is $0.64. Now find the percent increase.

Amount of increase = $p\%$ · Original wage   → $0.64 = p\%$ · $12.80

$$0.64 = p\% \cdot 12.80$$
$$\frac{0.64}{12.80} = \frac{p\% \cdot 12.80}{12.80}$$
$$0.05 = p\%$$
$$5\% = p$$

The percent increase in her hourly wage was 5%.

PREALGEBRA REVIEW

CHAPTER 6 PERCENTS

EXAMPLE 7 Through diet and exercise, Gabe's weight dropped from 236 pounds to 177 pounds. What was the percent decrease in his weight?

SOLUTION Gabe's original weight and his new weight are given. To find the percent decrease, first find the amount of decrease by subtracting.

Original weight − New weight   →   236 lb − 177 lb

$236 - 177 = 59$

The amount of decrease is 59 pounds. Now find the percent decrease.

Amount of decrease = $p\%$ · Original weight   → $59 = p\% \cdot 236$

$$59 = p\% \cdot 236$$
$$\frac{59}{236} = \frac{p\% \cdot 236}{236}$$
$$0.25 = p\%$$
$$25\% = p$$

The percent decrease in his weight was 25%.

EXAMPLE 8 During a clearance sale, a store gives 40% off the original price on all winter coats. What is the sale price of a coat that had an original price of $78.00?

SOLUTION In this problem the percent decrease and the original value are given. To find the sale price, first find the amount of decrease.

Amount of decrease = $p\%$ · Original price $a = 40\% \cdot \$78.00$

$a = 0.4 \cdot 78.00 = 31.20$

The amount of decrease was $31.20. To find the sale price, subtract.

Original price − Amount of decrease   → $78.00 − $31.20

$78.00 - 31.20 = 46.80$

The sale price of the coat is $46.80.

CHAPTER 6   PERCENTS   PREALGEBRA REVIEW

Simple Interest

Interest is an amount paid for borrowing money or the amount earned on an investment. The amount borrowed or invested is called the principal and the amount of interest is a percentage, called the interest rate, of the principal. **Simple interest** is interest that is calculated on the principal over time. The equation $I = P \cdot r \cdot t$, where I is the interest, P is the principal, r is the annual interest rate, and t is time (in years) can be used to calculate the simple interest.

EXAMPLE 9   What is the simple interest on an investment of $3500 at a simple interest rate of 3% for two years?

SOLUTION   The amount invested (the principal), the interest rate, and the time are given. To find the interest, use the simple interest equation.

$$I = P \cdot r \cdot t \quad \rightarrow \quad I = \$3500 \cdot 3\% \cdot 2$$

$$I = 3500 \cdot 0.03 \cdot 2 = 210$$

The simple interest is $210.

EXAMPLE 10   Tony borrows $2000 at an annual simple interest rate of $8\frac{1}{2}\%$ for 6 months. How much simple interest will he pay to the bank?

SOLUTION   The amount borrowed (the principal), the interest rate, and the time are given. Note that in the simple interest equation, time must be given in years. In this problem time is given in months. To find the amount of time in years, divide: 6 months $= \dfrac{6 \text{ months}}{12 \text{ months}} = \dfrac{1}{2}$ year. Now use the simple interest equation.

$$I = P \cdot r \cdot t \quad \rightarrow \quad I = \$2000 \cdot 8\frac{1}{2}\% \cdot 2 \quad \rightarrow \quad I = \$2000 \cdot 8.5\% \cdot \frac{1}{2}$$

$$I = 2000 \cdot 0.085 \cdot \frac{1}{2} = 85$$

Tony will pay simple interest of $85 to the bank.

PREALGEBRA REVIEW   CHAPTER 6   PERCENTS

Section 5: Exercises

Choose the correct answer.

1. A lab technician has 360 milliliters of a solution containing alcohol and water. If 15% of the solution is alcohol, how many milliliters are alcohol?

    (A) 40   (B) 54   (C) 38   (D) 48   (E) 50

2. On a particular day, 16% of the students at a high school were absent. If the school has 1250 students, how many were absent that day?

    (A) 150   (B) 250   (C) 125   (D) 200   (E) 300

3. The dinner bill at a restaurant for three people was $39.26. How much should be left for a 20% tip? Round to the nearest cent.

    (A) $7.85   (B) $7.52   (C) $7.80   (D) $6.90   (E) $7.86

4. Yi's rent is $480 per month, which is 24% of her monthly take-home salary. What is her monthly take home salary?

    (A) 1800   (B) $2000   (C) $115.20   (D) $2480   (E) $2200

5. At a small company, 28 of the 56 employees are women. What percent of the employees are women?

    (A) 40%   (B) 45%   (C) 50%   (D) 55%   (E) 60%

6. During an expansion, a ball park increased its seating capacity from 9500 to 10,450. What percent of the original seating capacity is the new seating capacity?

    (A) 10%   (B) 90%   (C) 105%   (D) 110%   (E) 115%

7. A publishing company shipped 60% of its inventory of a textbook. If 3900 textbooks were shipped, how many textbooks were in the original inventory?

    (A) 6000   (B) 6240   (C) 7000   (D) 6420   (E) 6500

8. The tank for a home's heating oil is 36% full. If there are 97.2 gallons of oil in the tank, what is the total capacity of tank?

    (A) 270 gal   (B) 150 gal   (C) 350 gal   (D) 194 gal   (E) 280 gal

CHAPTER 6   PERCENTS

9. An ATM charges customers a $2.50 fee for a withdrawal. What percent of a $100 withdrawal is the fee?

   (A) 25%   (B) 0.25%   (C) 2.5%   (D) 20.5%   (E) 250%

10. One serving of peanut butter contains 190 calories. If 140 of the calories are from fat, what percent of the calories are from fat? Round the answer to the nearest percent.

    (A) 65%   (B) 71%   (C) 26%   (D) 73%   (E) 74%

11. A real estate agent makes a $4\frac{1}{2}$% commission on the selling price of a house. What was the selling price of a home if the agent received a commission of $12,780?

    (A) 289,000   (B) $319,500   (C) $284,000   (D) $575.10   (E) $57,510

12. A student answered 85% of the questions on an exam correctly. If she answered 34 of the questions correctly, how many questions were on the exam?

    (A) 40   (B) 29   (C) 34   (D) 50   (E) 45

13. A basketball player makes 54 out of 72 field goal attempts. What percent of the field goal attempts did he miss?

    (A) 33%   (B) 75%   (C) 80%   (D) 25%   (E) 30%

14. A lab technician wants to make a solution of alcohol and water. How many milliliters of water must she add to 300 milliliters of alcohol so that the solution is 40% alcohol?

    (A) 450   (B) 40   (C) 200   (D) 750   (E) 350

15. A customer paid $2.97 in sales tax on a $36.00 purchase. What was the sales tax rate?

    (A) 12%   (B) 8.25%   (C) 7.5%   (D) 8%   (E) 9%

16. A customer paid $101.25 in sales tax on the purchase of a new plasma television. This amount was exactly $7\frac{1}{2}$% of the price of the television. What was the price of the television?

    (A) $1446   (B) $1250   (C) $1350   (D) $1500   (E) $1125

PREALGEBRA REVIEW																					CHAPTER 6 PERCENTS

17. There are 12 boys in a third grade class. This represents exactly 60% of the students in the class. How many girls are in the class?

   (A) 20   (B) 8   (C) 12   (D) 6   (E) 10

18. A student organization buys t-shirts for $12 each. What is the percent profit if they sell them for $16 each?

   (A) 40%   (B) 75%   (C) 25%   (D) $33\frac{1}{3}$%   (E) 35%

19. Last year 2200 students were enrolled at a small community college. This year the enrollment is 2376. What was the percent increase in the enrollment?

   (A) 90%   (B) 7%   (C) 9%   (D) 11%   (E) 8%

20. During a sale, the price of a sweater was decreased by $18. If the original price of the sweater was $60, what was the percent decrease in the price?

   (A) 70%   (B) 20%   (C) 30%   (D) 60%   (E) 18%

21. The area of a square is decreased from 160 square inches to 28.8 square inches. What percent decrease is this?

   (A) 85%   (B) 82%   (C) 18%   (D) 28.8%   (E) 78%

22. Claudio's annual salary is increased from $31,400 to $32,813. Find the percent increase.

   (A) 3%   (B) 4.3%   (C) 4.5%   (D) 3.5%   (E) 45%

23. Last year a magazine had 1,158,900 subscribers. This year the number of subscribers fell by 9%. How many subscribers does the magazine have this year?

   (A) 104,301   (B) 1,045,599   (C) 1,030,134   (D) 1,054,599   (E) 1,263,201

24. A jeweler marked up the price of a $160 ring by 40% to get the selling price. What was the selling price of the ring?

   (A) $200   (B) $224   (C) $96   (D) $64   (E) $196

CHAPTER 6   PERCENTS

25. During an expansion, a mall increased the number of parking spaces in its lot by 39%. How many parking spaces does the mall have after the expansion if 975 new parking spaces were added?

   (A) 2500   (B) 1355   (C) 3475   (D) 1525   (E) 3500

26. A customer bought a shirt that was marked 60% off the original price. What was the sale price of the shirt if the original price was $38?

   (A) $60.80   (B) $15.20   (C) $18.00   (D) $22.80   (E) $20.20

27. Mirna borrowed $1240 at an annual simple interest rate of $11\frac{1}{4}\%$ for one year. Find the interest paid on the loan.

   (A) $136.40   (B) $142.60   (C) $124.00   (D) $110.22   (E) $139.50

28. Find the interest earned on an investment of $4500 at an annual simple interest rate of 2% for 8 months.

   (A) $90.00   (B) $720.00   (C) $60.00   (D) $67.50   (E) $135.00

29. A deposit of $360 is made into an account at an annual simple interest rate of 5.5%. Assuming no additional deposits are made, how much is in the account in four years?

   (A) $417.60   (B) $439.20   (C) $376.20   (D) $406.80   (E) $366.48

30. Rajit takes out a loan for $8000 at an annual simple interest rate of 8.2% for five years. What is the total amount he pays back to the bank at the end of the loan period?

   (A) $3280   (B) $11,280   (C) $8656   (D) $40,800   (E) $11,200

31. The length of a rectangular garden is 12 feet and the width is 16 feet. If the length and width are increased by 25%, by what percent is the area of the garden increased?

   (A) 56.25%   (B) 25%   (C) 50%   (D) 64%   (E) 156.25%

32. Suppose a loan of $1000 is taken out at an annual simple interest rate of 9% for three years. What percent of the principal is the interest paid on the loan?

   (A) 9%   (B) 18%   (C) 27%   (D) 15%   (E) 21%

33. An apartment building has studio, one bedroom, and two bedroom apartments. There are a total of 50 apartments in the building. 24% of the apartments are studios. There are 10 more one bedroom apartments than studio apartments. What percent of the apartments in the building are two bedroom apartments?

 (A) 46%   (B) 56%   (C) 76%   (D) 68%   (E) 32%

34. A town had a fiscal year budget of $72,000,000. 48% of the town's fiscal year budget is allocated for the school department. If 36% of the amount allocated for the school department is used for classroom teachers, what percent of the town's budget is used for classroom teachers?

 (A) $17.28%   (B) 36%   (C) 18.25%   (D) 16%   (E) 84%

35. In a chemistry class of 80 students, 55% are chemistry majors. Of the remaining students, 75% are biology majors. What percent of the students are not chemistry or biology majors?

 (A) $33\frac{3}{4}$%   (B) $41\frac{1}{4}$%   (C) 8%   (D) $11\frac{1}{4}$%   (E) $10\frac{3}{4}$%

36. During a taste test, 125 people were asked which brand they preferred, Brand A, Brand B, or Brand C. Two-fifths of the people preferred Brand A. Of the remaining people surveyed, 60% preferred Brand B. What percent preferred Brand C?

 (A) 20%   (B) 28%   (C) 24%   (D) 36%   (E) 32%

37. There are 36 students in a math class. Of the 16 women in the class, 75% are enrolled in the nursing program. Of the 20 men in the class, 15% are enrolled in the nursing program. What percent of the total number of students in the class are enrolled in the nursing program?

 (A) 90%   (B) $40\frac{1}{2}$%   (C) $33\frac{2}{3}$%   (D) $41\frac{2}{3}$%   (E) 45%

38. A company has 150 employees, of which 38% work part-time. Assuming there are no layoffs, how many more part-time employees must be hired so that 50% of the employees work part-time?

 (A) 36   (B) 57   (C) 18   (D) 93   (E) 24

CHAPTER 6   PERCENTS

39. The asking price of a house is $220,000. After 3 months on the market, the owners reduced the price by $10,000. After 6 months on the market, they reduce the price by 10%. What was the final asking price of the home?

(A) $210,000   (B) $189,000   (C) $198,000   (D) $200,000   (E) $188,000

40. A pair of jeans is marked 40% off the original price of $84. In addition, the store gives an additional 20% off the sale price. What is the final sale price of the pair of jeans?

(A) $33.60   (B) $61.60   (C) $40.32   (D) $6.72   (E) $26.88

41. A DVD player that originally cost the store $64 was marked up 50%. During a sale, the price was marked down 25%. What was the sale price of the DVD player?

(A) $64   (B) $32   (C) $72   (D) $96   (E) $80

42. In the year 2000, the population of a town was 12,800. By 2003, the population grew 5%. It grew another 5% from 2003 to 2006. What was the overall percent increase in the population from 2000 to 2006?

(A) 10%   (B) 5.25%   (C) 9%   (D) 10.25%   (E) 5%

PREALGEBRA REVIEW

CHAPTER 6   PERCENTS

Section 6: Mean, Median, and Mode

Statistics is a branch of mathematics in which sets of numbers called data are organized and studied. The mean, the median, and the mode are three statistics used to describe a set of data. These numbers are used as *center points* or *measures of central tendency*.

Mean

The **mean**, or *average* of a set of data is the sum of the values in the set of data divided by the number of values in the data set. (Recall that averages were discussed in Chapter 1.)

EXAMPLE 1   A golfer had scores of 72, 65, 68, and 67 in the four rounds of a tournament. Find her mean score per round.

SOLUTION   The scores for each round are given. To find the mean, add the scores and divide by the number of values in the data set.

$$\frac{\text{Sum of scores}}{\text{Number of scores}} = \frac{72+65+68+67}{4} = \frac{272}{4} = 68$$

Her mean score per round was 68.

Median

Very high or very low values can affect the mean. Because of this, it is not always a good indicator of the central tendency of the data. In this case, the **median** provides a better measure of central tendency.

To find the median, write the values in order from smallest to largest. If the set of data has an odd number of values, the median is the middle value.

EXAMPLE 2   Find the median of the set of numbers: 12, 3, 8, 11, 5, 1, 9

SOLUTION   First arrange the numbers in order from smallest to largest: 1, 3, 5, 8, 9, 11, 12
Since there are an odd number of values, the median is the middle value.

1, 3, 5, **8**, 9, 11, 12
↑
Middle
value

CHAPTER 6   PERCENTS

Some data sets have an even number of values. In this case, there is no middle value. To find the median, find the mean of the middle two values.

EXAMPLE 3   What is the median of the following test scores?

53, 89, 97, 64, 78, 76, 32, 99

SOLUTION   Arrange the numbers in order from smallest to largest.

32, 54, 64, $\underbrace{76, 78}_{\text{Middle two values}}$, 89, 97, 99

The middle two values are 76 and 78. Now find the mean of these two values.

$$\frac{76+78}{2} = \frac{154}{2} = 77$$

The median is 77.

Mode

The **mode** is the value or values that occur most often in a set of data. If two values occur the same number of time, the set of data has two modes and is called *bimodal*. If each value occurs the same number of times, then there is no mode. Note that although it would be helpful, the values do not have to be arranged in order from smallest to largest to find the mode.

EXAMPLE 4   Find the mode of the set of data: 34, 31, 45, 30, 31, 36, 31, 30

SOLUTION   The mode is 31 since it occurs three times, which is more often than any other number in the set of data.

EXAMPLE 5   The birth lengths (in inches) of babies born at a hospital this week were 19, 22, 21, 19, 21, 23.

SOLUTION   The values 19 and 21 both occur twice, therefore each is a mode of the set of data. This set of data is bimodal.

Section 6: Exercises

Find the mean.

1. 126, 109, 144, 95, 137, 118, 139

2. 0.15, 0.36, 0.24, 0.5, 0.49, 0.84

3. Kay had test scores of 92, 86, and 80.

4. The daily high temperatures this week were 69°, 75°, 78°, 71°, 80°, 75°, 77°.

5. Renee received the following number of e-mails over the last five days: 16, 23, 11, 17, and 8.

6. Don's monthly phone bills for the last four months were $68.34, $79.17, $61.53, and $76.12.

Find the median.

7. 17, 9, 20, 4, 15, 12, 18, 35, 16

8. 2006, 1891, 1607, 1957, 1971, 1492

9. The ages of instructors in the math department are 54, 32, 29, 42, 47, and 38.

10. A basketball team's scores for the last five home games were 98, 106, 94, 112, and 89.

11. The number of hours a student studied this week were 2.5, 5, 6, 4, 3, 6, and 8.

12. The monthly rainfall (in inches) for each month from July through December was 3, 1, 4, 3, 2, and 5.

Find the mode.

13. 107, 132, 98, 170, 132, 89, 116

14. 7.8, 5.9, 6.5, 9.7, 5.9, 7.2, 5.9, 7.8

15. Mary's annual percent pay increases for the last four years were 5.9%, 4.8%, 6%, and 4.3%

16. A student's quiz scores were 89, 96, 84, 96, 86, 78, and 89.

17. The miles an athlete ran each day last week were 6.5, 8, 8, 6, 7.5, and 7.

18. Ursula's scores for the last eight games she bowled were 156, 163, 158, 171, 149, 156, 165, and 160.

PREALGEBRA REVIEW   PRACTICE TESTS

Practice Test A

1. Simplify: $6^2 - 18 \div 3 \times 6 + 23$

   (A) 59 (B) 23 (C) −1 (D) 58 (E) 34

2. Evaluate: $\left(\dfrac{1}{2} - \dfrac{3}{8}\right) \div \left(\dfrac{5}{6} - \dfrac{1}{3}\right)$

   (A) $\dfrac{1}{8}$ (B) $\dfrac{1}{4}$ (C) $\dfrac{1}{16}$ (D) $\dfrac{3}{4}$ (E) 1

3. Evaluate: $10 - 3(-4)$

   (A) −28 (B) −2 (C) 11 (D) 22 (E) 2

4. Bill bought 100 baseball caps for $8 each. The manufacturer gives him another 10 caps at no charge. What is his total profit if he sold all the caps for $12 each?

   (A) $800 (B) $1320 (C) $520 (D) $20 (E) $400

5. A taxi company charges $1.75 for the first $\dfrac{1}{2}$ mile and $0.25 for each additional $\dfrac{1}{8}$ mile for a ride. In addition, the company charges a $3.00 fee for rides originating at the airport. How much would a twelve mile taxi ride to the airport cost?

   (A) $24.75 (B) $23.00 (C) $27.75 (D) $24.00 (E) $60.00

6. The product of 1.6×10^7 and 2×10^{-3} in standard form is

   (A) 3200 (B) 32,000 (C) 0.032 (D) 3.2 (E) 32,000,000,000

7. If $1 < \sqrt{x} < 4$, then which of the following must be true about x?

   (A) $2 < x < 8$ (B) $1 < x < 2$ (C) $4 < x < 16$ (D) $1 < x < 4$ (E) $1 < x < 16$

8. Find the value of x: $\dfrac{0.7}{2} = \dfrac{21}{x}$

   (A) 60 (B) 42 (C) 6 (D) 30 (E) 14

9. Evaluate: $\dfrac{(2^3)^2}{16}$

   (A) $\dfrac{3}{4}$ (B) $\dfrac{1}{2}$ (C) $\dfrac{9}{4}$ (D) 2   (E) 4

10. Linda pays $63 every three months for life insurance. At this rate, what is the total amount she will pay for life insurance in $4\dfrac{1}{2}$ years?

    (A) $21.00   (B) $283.50   (C) $1134.00   (D) $1008.00   (E) $225.00

11. What is 3.5% of 70?

    (A) 245   (B) 2.45   (C) 0.245   (D) 2000   (E) 24.5

12. Evaluate $2x^2 - 5y^2$ for $x = -0.5$ and $y = -1$.

    (A) 7   (B) 4.5   (C) –4   (D) –4.5   (E) 5.5

13. A math test was given to 40 students. The girls got an average score of 75, whereas the boys got an average score of 70. What was the average score for all 40 students, if 16 girls and 24 boys took the test?

    (A) 72   (B) 70   (C) 74   (D) 75   (E) 73

14. Wendy went on a diet to lose weight. She lost 8 pounds the first month, 7 pounds the second month, 4 pounds the third month and 5 pounds the fourth month. What was her mean weight loss each month?

    (A) 5 lb   (B) 4 lb   (C) 24 lb   (D) 8 lb   (E) 6 lb

15. A school had 1800 students. 30% were freshman and 24% were sophomores. There were 54 fewer juniors than sophomores. What percent of the students were seniors?

    (A) 46%   (B) 30%   (C) 25%   (D) 19%   (E) 20%

PREALGEBRA REVIEW   PRACTICE TESTS

Practice Test B

1. $16 + 24 \div 8 \times 3 - 5 =$

   (A) 20   (B) 10   (C) 12   (D) 52   (E) 5

2. When $\left(\dfrac{1}{4}+\dfrac{1}{3}\right)-\left(\dfrac{1}{2}-\dfrac{1}{6}\right)$ is calculated and simplified, the denominator of the resulting fraction is

   (A) 6   (B) 12   (C) 4   (D) 3   (E) 2

3. Evaluate: $8(-6) \div 12 + 4(-1-3)^2$

   (A) 0   (B) 60   (C) –36   (D) 64   (E) 12

4. Carrie bought $2\dfrac{1}{2}$ pounds of apples, 3.68 pounds of bananas, and $1\dfrac{3}{5}$ pounds of peaches. How many pounds of fruit did she buy?

   (A) 7.78   (B) 7.08   (C) 7.38   (D) 7.80   (E) 7.33

5. A calling card service charges a connection fee of $0.35 for each call plus an additional $0.03 per minute. Julia made 12 calls that totaled 9.6 hours. How much money does she have left on a $50.00 calling card?

   (A) $25.40   (B) $21.48   (C) $17.28   (D) $28.52   (E) $4.49

6. What is the perimeter of a square if its area is 121 square centimeters?

   (A) 11 cm   (B) 40 cm   (C) 44 cm   (D) 22 cm   (E) 48 cm

7. Calculate: $\dfrac{7.2 \times 10^7}{(1.5 \times 10^5)(3 \times 10^{-12})}$

   (A) 1.6×10^0 (B) 1.6×10^{14} (C) 1.6×10^{-10} (D) 1.6×10^7 (E) 1.6×10^{24}

8. What percent of 52 is 13?

   (A) 20%   (B) 15%   (C) 25%   (D) 40%   (E) 24%

PRACTICE TESTS   PREALGEBRA REVIEW

9. What is the value of n in the proportion $\dfrac{n}{16} = \dfrac{5}{12}$?

   (A) $5\dfrac{3}{5}$ (B) $6\dfrac{2}{3}$ (C) $1\dfrac{3}{4}$ (D) $38\dfrac{2}{5}$ (E) 6

10. What is the value of $a^2 - b^2$ when $a = 9$ and $b = \sqrt{3}$?

    (A) $\sqrt{6}$ (B) 15   (C) 78   (D) 36   (E) 72

11. A baseball team won 57 out of 76 games played. Write a ratio of the games lost to the games played in simplest form.

    (A) $\dfrac{1}{4}$ (B) $\dfrac{4}{3}$ (C) $\dfrac{1}{3}$ (D) $\dfrac{3}{4}$ (E) $\dfrac{2}{3}$

12. A car uses $\dfrac{1}{2}$ gallon of gasoline for every 14.2 miles it is driven. How many gallons of gasoline will it use if it is driven 14,484 miles?

    (A) 2040   (B) 510   (C) 540   (D) 570   (E) 1020

13. Eighty of the members of a student organization are men. This represents exactly 40% of the members. How many members are women?

    (A) 200   (B) 32   (C) 120   (D) 48   (E) 40

14. Adams, Davis, Jackson, and Pisoli, ran for election to the school committee. Adams received 36% of the votes and Pisoli received 65% of the remaining votes. Jackson received one-third the number of votes Adams received. If 8250 votes were cast, what percent of the votes did Davis receive?

    (A) 25%   (B) 41.6%   (C) 22.4%   (D) 10.4%   (E) 16.25%

15. Maura had an average of 88 on her first three exams of the semester. What must she receive on her next exam so that her average for all four exams is 90?

    (A) 98   (B) 96   (C) 93   (D) 90   (E) 95

PREALGEBRA REVIEW   PRACTICE TESTS

Practice Test C

1. Calculate: $120 \div (7-5)^3 \cdot 3 + 45$

    (A) 50   (B) 45   (C) 90   (D) 225   (E) 135

2. $\dfrac{3}{5} - \dfrac{4}{9}\left(\dfrac{2}{3} - \dfrac{1}{6}\right) \div \dfrac{5}{6} =$

    (A) $\dfrac{1}{3}$ (B) $\dfrac{1}{2}$ (C) $\dfrac{4}{15}$ (D) $\dfrac{3}{5}$ (E) $\dfrac{1}{15}$

3. Evaluate: $1.8 + 1.5 \div 0.3 - (0.5)^2$

    (A) 8.5   (B) 11.25   (C) 4.3   (D) 2.55   (E) 6.55

4. Eight students are planning to go skiing for the day. Each lift ticket costs $54 or $38 for groups of 10 or more. How much would the eight students save if they get two more people to go with them and they get lift tickets at the group rate?

    (A) $16   (B) $160   (C) $128   (D) $64   (E) $96

5. $2\dfrac{1}{2} + 3.6 \times \dfrac{8}{9} - 4.8 =$

    (A) 0.2   (B) 0.9   (C) 0.5   (D) 0.3   (E) 0.6

6. Summer makes bars of glycerin soap. It costs her $1.15 per bar for supplies and she sells each bar for $2.75. How many bars of soap must she sell in order to make a profit of $88?

    (A) 55   (B) 44   (C) 32   (D) 36   (E) 60

7. Calculate $2,400,000,000 \times 0.00003$ and write the answer in scientific notation.

    (A) 7.2×10^{13} (B) 7.2×10^9 (C) 7.2×10^5 (D) 7.2×10^4 (E) 7.2×10^{14}

8. Which of the following statements is true?

    (A) $4 < \sqrt{7} < 16$ (B) $6 < \sqrt{7} < 8$ (C) $36 < \sqrt{7} < 64$
    (D) $2 < \sqrt{7} < 3$ (E) $4 < \sqrt{7} < 8$

9. Solve for n: $\dfrac{8}{n} = \dfrac{6}{7}$

(A) 2 (B) $9\dfrac{1}{3}$ (C) $2\dfrac{1}{2}$ (D) 9 (E) $9\dfrac{2}{3}$

10. 1.8 is 0.6% of what number?

(A) 30 (B) 0.0108 (C) 3 (D) 1.08 (E) 300

11. During a snowstorm, 0.25 inch of snow fell every 20 minutes. At this rate, how many inches of snow fell in 4 hours?

(A) 10 (B) 0.1 (C) 3 (D) 6 (E) 4

12. A book shop had sales of $1234.68, $896.59, $521.61, $744.09, $900.77, and $2486.46 over the past six days. Find the median sales over the past six days.

(A) $898.68 (B) $896.59 (C) $1130.70 (D) $632.85 (E) $900.77

13. Evaluate $x(2x-3y)^2$ for $x = -2$ and $y = -3$.

(A) 50 (B) –72 (C) –20 (D) 52 (E) –50

14. An exam had 15 fill in the blank questions, 20 multiple choice questions, and 10 short answer questions. A student answered 80% of the fill in the blank questions, 85% of the multiple choice questions, and 70% of the short answer questions correctly. What percent of the exam questions did the student answer incorrectly?

(A) 80% (B) 75% (C) 20% (D) 30% (E) 22%

15. Lou had an average of 84 on his first 6 quizzes in his algebra class. He got scores of 88, 97, 80, 85, 98, and 92 on his next six quizzes. What was his average for all twelve quizzes?

(A) 85 (B) 89 (C) 84 (D) 87 (E) 86

A. Answers to Exercises

Chapter 1

Section 1
1. 4 **2.** 8 **3.** 2 **4.** 6 **5.** 1 ten thousand + 2 thousands + 3 hundreds + 0 tens + 7 ones **6.** 2 millions + 3 hundred thousands + 4 ten thousands + 5 thousands + 6 hundreds + 7 tens + 7 ones **7.** 4 thousands + 0 hundreds + 3 tens + 1 one **8.** 4 hundred millions + 5 ten millions + 6 millions + 7 hundred thousands + 8 ten thousands + 3 thousands + 2 hundreds + 0 tens + 9 ones **9.** 38,061 **10.** 5,991,426 **11.** 850 **12.** 611,494,116 **13.** two hundred twenty-seven thousand, eight hundred ninety-six **14.** two hundred fifty-six **15.** nine thousand seven hundred five **16.** thirteen million, five hundred eight thousand, seven hundred forty-three **17.** 843 **18.** 9867 **19.** 15,495 **20.** 11,639 **21.** 1601 **22.** 16,713 **23.** 578 **24.** 1115 **25.** 115,679 **26.** 77,619 **27.** 748 **28.** 2350 **29.** 45,976 **30.** 46,878 **31.** 150,528 **32.** 941,334 **33.** 168 R2 **34.** 952 R3 **35.** 65 **36.** 264 R16 **37.** 175 R184 **38.** 3209 R170 **39.** B **40.** A **41.** A **42.** E **43.** C **44.** B **45.** E **46.** C **47.** A **48.** D **49.** B **50.** A **51.** A **52.** C **53.** E **54.** B **55.** C **56.** D **57.** A **58.** C

Section 2
1. +6194 meters, or 6194 meters **2.** −4 **3.** −43 pounds **4.** +8 yards, or 8 yards **5.** > **6.** < **7.** < **8.** > **9.** −19, −9, −8, −1, 2, 6, 7, 11 **10.** −14, −11, −7, −4, 0, 4, 9, 16 **11.** 8 **12.** 5 **13.** 2 **14.** 17 **15.** > **16.** = **17.** < **18.** < **19.** −11 **20.** 9 **21.** 14 **22.** 0 **23.** 10 **24.** −8 **25.** −5 **26.** −13

Section 3
1. 6 **2.** −19 **3.** −9 **4.** −1 **5.** −72 **6.** −35 **7.** 18 **8.** 0 **9.** 9 **10.** 119 **11.** −173 **12.** 0 **13.** −11 **14.** −1 **15.** −11 **16.** 7 **17.** −27 **18.** −72 **19.** −4 **20.** −12 **21.** 22 **22.** 13 **23.** 135 **24.** 88 **25.** 0 **26.** 67 **27.** −3 **28.** −12 **29.** −36 **30.** 28 **31.** −1200 **32.** 396 **33.** B **34.** C **35.** A **36.** D **37.** B **38.** C **39.** D **40.** B **41.** C **42.** A **43.** B **44.** B **45.** C **46.** B **47.** C **48.** A **49.** E **50.** E

Section 4
1. 6^3 **2.** 9^5 **3.** $(-7)^4$ **4.** $(-12)^2$ **5.** $8^4 \cdot 3$ **6.** $5 \cdot (-2)^2$ **7.** 16 **8.** 125 **9.** −1 **10.** 81 **11.** −16 **12.** −64 **13.** 10,000 **14.** 100,000 **15.** 16 **16.** 57 **17.** 24 **18.** 28 **19.** 8 **20.** 18 **21.** 44 **22.** 29 **23.** 17 **24.** −600 **25.** 400 **26.** 81 **27.** 25 **28.** 23 **29.** 56 **30.** 0 **31.** 9 **32.** 20 **33.** 10 **34.** 75 **35.** −21 **36.** 3 **37.** −1 **38.** 2

Section 5
1. 4 **2.** 10 **3.** 23 **4.** 39 **5.** −2 **6.** −6 **7.** C **8.** E **9.** A **10.** C **11.** A **12.** D **13.** B **14.** A **15.** B **16.** A **17.** B **18.** D **19.** E **20.** B **21.** A **22.** D **23.** C **24.** E **25.** B **26.** D

Chapter 2

Section 1
1. 1, 2, 5, 10 **2.** 1, 7 **3.** 1, 2, 3, 6, 9, 18 **4.** 1, 2, 4, 8, 16, 32 **5.** 1, 5, 17, 85 **6.** 1, 2, 3, 4, 6, 9, 12, 18, 27, 36, 54, 108 **7.** composite number **8.** prime number **9.** prime number **10.** composite number **11.** $2 \cdot 2 \cdot 2 \cdot 2$ **12.** $3 \cdot 13$ **13.** $3 \cdot 3 \cdot 7$ **14.** $2 \cdot 2 \cdot 2 \cdot 2 \cdot 5$ **15.** 16 **16.** 6 **17.** 4 **18.** 5 **19.** 48 **20.** 11 **21.** 5, 10, 15, 20, 25, 30 **22.** 8, 16, 24, 32, 40, 48 **23.** 24 **24.** 30 **25.** 72 **26.** 56 **27.** 54 **28.** 210 **29.** 36 **30.** 60 **31.** C **32.** C **33.** A **34.** C **35.** B **36.** E **37.** D **38.** A

Section 2
1. $\frac{2}{4}$ **2.** $\frac{3}{8}$ **3.** $\frac{7}{10}$ **4.** $\frac{2}{2}=1$ **5.** $\frac{13}{9}$ **6.** $\frac{17}{6}$ **7.** improper fraction **8.** proper fraction **9.** improper fraction **10.** mixed number **11.** mixed number **12.** proper fraction **13.** $2\frac{2}{5}$ **14.** $8\frac{1}{2}$ **15.** $5\frac{6}{7}$ **16.** $-4\frac{5}{12}$ **17.** 6 **18.** $5\frac{1}{15}$ **19.** $-10\frac{1}{10}$ **20.** 4 **21.** $-5\frac{7}{8}$ **22.** −6 **23.** $8\frac{11}{12}$ **24.** $5\frac{13}{24}$ **25.** $\frac{7}{5}$ **26.** $\frac{13}{4}$ **27.** $-\frac{40}{7}$ **28.** $\frac{31}{11}$ **29.** $\frac{43}{10}$ **30.** $-\frac{53}{15}$

ANSWERS TO EXERCISES AND PRACTICE TESTS   PREALGEBRA REVIEW

Chapter 2, Section 2 (continued)

31. $\dfrac{124}{9}$ 32. $\dfrac{79}{20}$ 33. $\dfrac{203}{3}$ 34. $\dfrac{83}{8}$

35. $-\dfrac{82}{13}$ 36. $-\dfrac{95}{6}$ 37. $3\dfrac{3}{8}$ 38. $\dfrac{3}{4}$ 39. $\dfrac{6}{11}$

40. $4\dfrac{1}{2}$ 41. A   42. B   43. C   44. D   45. A

46. E   47. C   48. A   49. D   50. C

Section 3

1. $\dfrac{9}{18}$ 2. $\dfrac{14}{21}$ 3. $\dfrac{48}{60}$ 4. $\dfrac{18}{24}$ 5. $-\dfrac{16}{56}$

6. $\dfrac{-8}{48}$ 7. $\dfrac{36}{12}$ 8. $\dfrac{9}{9}$ 9. $\dfrac{28}{32}$ 10. $\dfrac{24}{54}$

11. $\dfrac{24}{72}$ 12. $\dfrac{30}{90}$ 13. $\dfrac{-56}{100}$ 14. $-\dfrac{66}{96}$

15. 12; $\dfrac{9}{12},\dfrac{8}{12}$ 16. 24; $\dfrac{3}{24},\dfrac{18}{24}$ 17. 45; $\dfrac{30}{45},\dfrac{40}{45}$

18. 48; $\dfrac{12}{48},\dfrac{40}{48}$ 19. 72; $\dfrac{36}{72},\dfrac{24}{72}$ 20. 54; $\dfrac{30}{54},\dfrac{39}{54}$

21. $\dfrac{2}{3}$ 22. $\dfrac{2}{5}$ 23. $\dfrac{2}{7}$ 24. $\dfrac{6}{7}$ 25. $-\dfrac{2}{3}$

26. $-\dfrac{3}{4}$ 27. $\dfrac{2}{3}$ 28. $\dfrac{1}{2}$ 29. $\dfrac{1}{2}$ 30. $\dfrac{3}{8}$

31. $\dfrac{-4}{7}$ 32. $\dfrac{-3}{4}$ 33. cannot be simplified further   34. cannot be simplified further   35. $\dfrac{-7}{10}$

36. $-\dfrac{16}{25}$ 37. >   38. <   39. <   40. >   41. >

42. <   43. >   44. >   45. C   46. A   47. B

48. D   49. C   50. E   51. C   52. A   53. B

54. A   55. C   56. D   57. E   58. A

Section 4

1. $\dfrac{5}{4}=1\dfrac{1}{4}$ 2. $\dfrac{10}{11}$ 3. $\dfrac{43}{30}=1\dfrac{13}{30}$

4. $\dfrac{17}{12}=1\dfrac{5}{12}$ 5. $\dfrac{11}{8}=1\dfrac{3}{8}$ 6. $\dfrac{8}{9}$ 7. $-\dfrac{19}{20}$

8. $\dfrac{16}{35}$ 9. $\dfrac{95}{72}=1\dfrac{23}{72}$ 10. $\dfrac{73}{48}=1\dfrac{25}{48}$

11. $-\dfrac{4}{45}$ 12. $-\dfrac{2}{3}$ 13. $\dfrac{3}{5}$ 14. $\dfrac{1}{5}$ 15. $\dfrac{7}{20}$

16. $\dfrac{5}{21}$ 17. $-\dfrac{1}{16}$ 18. $\dfrac{8}{27}$ 19. $\dfrac{7}{48}$

20. $\dfrac{7}{6}=1\dfrac{1}{6}$ 21. $\dfrac{23}{72}$ 22. $\dfrac{19}{144}$

23. $-\dfrac{19}{18}=-1\dfrac{1}{18}$ 24. $-\dfrac{21}{22}$ 25. $4\dfrac{1}{5}$ 26. $3\dfrac{2}{3}$

27. $1\dfrac{11}{12}$ 28. $1\dfrac{7}{24}$ 29. $-1\dfrac{7}{15}$ 30. $-1\dfrac{4}{9}$

31. $\dfrac{23}{24}$ 32. $9\dfrac{9}{10}$ 33. $-2\dfrac{1}{4}$ 34. $-13\dfrac{4}{5}$

35. $5\dfrac{19}{24}$ 36. $7\dfrac{11}{14}$ 37. $-\dfrac{64}{75}$ 38. $2\dfrac{7}{15}$

39. $-16\dfrac{1}{2}$ 40. $-6\dfrac{1}{4}$ 41. $-3\dfrac{1}{3}$ 42. $2\dfrac{1}{2}$

43. $-\dfrac{5}{6}$ 44. $-18\dfrac{2}{3}$ 45. $\dfrac{23}{12}=1\dfrac{11}{12}$

46. $\dfrac{79}{40}=1\dfrac{39}{40}$ 47. $5\dfrac{13}{18}$ 48. $4\dfrac{5}{24}$ 49. $-1\dfrac{4}{21}$

50. $4\dfrac{13}{36}$ 51. B   52. E   53. A   54. D   55. A

56. C   57. A   58. D   59. D   60. A   61. B

62. B   63. E   64. C   65. B   66. E   67. B

68. D   69. C   70. D   71. D   72. B   73. A

74. E

Section 5

1. $\dfrac{1}{20}$ 2. $\dfrac{4}{15}$ 3. $\dfrac{5}{22}$ 4. $\dfrac{5}{6}$ 5. $\dfrac{1}{5}$ 6. $\dfrac{3}{7}$

7. $-\dfrac{1}{12}$ 8. $\dfrac{16}{39}$ 9. $\dfrac{9}{4}=2\dfrac{1}{4}$ 10. $\dfrac{14}{3}=4\dfrac{2}{3}$

11. $\dfrac{10}{9}=1\dfrac{1}{9}$ 12. $-\dfrac{26}{7}=-3\dfrac{5}{7}$ 13. $\dfrac{3}{2}=1\dfrac{1}{2}$

14. $\dfrac{4}{5}$ 15. -160 16. -69 17. 33   18. 16

19. -50 20. $\dfrac{29}{7}=4\dfrac{1}{7}$ 21. $\dfrac{3}{1}=3$ 22. $\dfrac{11}{8}$

23. $\dfrac{9}{16}$ 24. $\dfrac{4}{15}$ 25. $\dfrac{1}{4}$ 26. $\dfrac{1}{2}$ 27. $-\dfrac{11}{10}$

28. $-\dfrac{12}{21}$ 29. $\dfrac{4}{5}$ 30. $\dfrac{9}{10}$ 31. $\dfrac{14}{15}$

32. $\dfrac{7}{5}=1\dfrac{2}{5}$ 33. $-\dfrac{16}{5}=-3\dfrac{1}{5}$ 34. $-\dfrac{3}{4}$

35. $\dfrac{7}{4}=1\dfrac{3}{4}$ 36. $\dfrac{2}{3}$ 37. $\dfrac{16}{3}=5\dfrac{1}{3}$ 38. $\dfrac{1}{12}$

PREALGEBRA REVIEW

Chapter 2, Section 5 (continued)

39. $\frac{1}{3}$ **40.** 2 **41.** –4 **42.** $\frac{1}{8}$ **43.** $\frac{3}{4}$
44. $\frac{12}{7} = 1\frac{5}{7}$ **45.** $\frac{14}{25}$ **46.** $-\frac{74}{45} = -1\frac{29}{45}$
47. $\frac{7}{3} = 2\frac{1}{3}$ **48.** $\frac{16}{5} = 3\frac{1}{5}$ **49.** D **50.** B
51. C **52.** A **53.** E **54.** C **55.** A **56.** C
57. E **58.** D **59.** B **60.** A **61.** C **62.** B
63. D **64.** E **65.** B **66.** C **67.** D **68.** B
69. D **70.** C **71.** A **72.** B **73.** D **74.** B
75. B **76.** E

Section 6

1. $\frac{1}{32}$ **2.** $\frac{1}{27}$ **3.** $-\frac{27}{64}$ **4.** $\frac{49}{81}$ **5.** $\frac{64}{9} = 7\frac{1}{9}$
6. $-\frac{125}{8} = -15\frac{5}{8}$ **7.** $\frac{7}{15}$ **8.** $-\frac{2}{3}$ **9.** $\frac{21}{4} = 5\frac{1}{4}$
10. $\frac{11}{30}$ **11.** 128 **12.** $\frac{1}{12}$ **13.** $\frac{47}{20} = 2\frac{7}{20}$
14. $-\frac{13}{12} = -1\frac{1}{12}$ **15.** $\frac{41}{48}$ **16.** $\frac{97}{36} = 2\frac{25}{36}$
17. $\frac{5}{24}$ **18.** $-\frac{35}{1044}$ **19.** $-\frac{19}{2} = -9\frac{1}{2}$
20. $-\frac{19}{10} = -1\frac{9}{10}$

Section 7

1. $\frac{2}{15}$ **2.** 2 **3.** $\frac{5}{2} = 2\frac{1}{2}$ **4.** $\frac{7}{108}$ **5.** $\frac{5}{4} = 1\frac{1}{4}$
6. $\frac{2}{3}$ **7.** $-\frac{16}{21}$ **8.** $\frac{80}{49} = 1\frac{31}{49}$ **9.** $-\frac{16}{3} = -5\frac{1}{3}$
10. $-\frac{3}{50}$ **11.** $\frac{2}{15}$ **12.** $\frac{25}{6} = 4\frac{1}{6}$ **13.** $\frac{44}{57}$
14. $\frac{147}{125} = 1\frac{22}{125}$ **15.** $\frac{3}{4}$ **16.** $\frac{10}{7} = 1\frac{3}{7}$ **17.** $\frac{4}{9}$
18. 3 **19.** 8 **20.** $\frac{27}{44}$ **21.** $\frac{105}{68} = 1\frac{37}{68}$
22. $\frac{21}{5} = 4\frac{1}{5}$ **23.** $-\frac{125}{12} = -10\frac{5}{12}$ **24.** $-\frac{213}{368}$

Chapter 3

Section 1

1. 5 **2.** 4 **3.** 7 **4.** 9 **5.** 1 **6.** 2
7. 3 tenths + 7 hundredths + 9 thousandths
8. 1 one + 2 tenths **9.** 9 tens + 4 ones + 7 tenths + 5 hundredths **10.** 4 ones + 0 tenths + 2 hundredths + 6 thousandths + 1 ten-thousandth + 4 hundred-thousandths **11.** 0.82 **12.** 9.076 **13.** 461.9083
14. 1500.9 **15.** seven and fifteen hundredths
16. four hundred thirty-one thousandths **17.** two hundred fifty and five tenths **18.** twenty-three and eight hundred seventy-two ten-thousandths
19. $\frac{4}{10}$ **20.** $\frac{567}{1000}$ **21.** $\frac{1}{100}$ **22.** $\frac{23}{10,000}$
23. $-\frac{233}{1000}$ **24.** $-\frac{9}{10,000}$ **25.** $\frac{165}{100} = 1\frac{65}{100}$
26. $\frac{216}{10} = 21\frac{6}{10}$ **27.** $-\frac{9425}{1000} = -9\frac{425}{1000}$
28. $-\frac{14,299}{100} = -142\frac{99}{100}$ **29.** $\frac{502}{100} = 5\frac{2}{100}$
30. $\frac{16,003}{1000} = 16\frac{3}{1000}$ **31.** 0.13 **32.** 0.2
33. 0.0451 **34.** 0.008 **35.** –0.45 **36.** –0.407
37. 3.07 **38.** 10.1 **39.** –12.067 **40.** 34.59
41. –1.0001 **42.** 200.8 **43.** D **44.** B **45.** A
46. D

Section 2

1. > **2.** < **3.** < **4.** < **5.** > **6.** >
7. 0.0589, 0.5, 0.5098, 0.55908, 0.598
8. 4.11, 4.113, 4.13, 14.31, 41.13 **9.** 0.99, 0.909, 0.099, 0.09, 0.009 **10.** 62.2, 60.2, 6.22, 0.602, 0.062 **11.** 0.63 **12.** 0.7021 **13.** 9.093
14. 5.5 **15.** > **16.** = **17.** < **18.** < **19.** >
20. > **21.** 2.79 **22.** 1.0 **23.** 1.006 **24.** 4.0
25. 8.00 **26.** 0.500 **27.** 5.9 **28.** 300 **29.** C
30. B **31.** A **32.** B **33.** E **34.** A

Section 3

1. 12.95 **2.** 1 **3.** 28.342 **4.** 4.45199 **5.** 3.066
6. 4.7081 **7.** 66.2152 **8.** 138.1807 **9.** 0.52
10. 6.639 **11.** 0.1032 **12.** 8.30145 **13.** 0.0001
14. 9.99 **15.** 22.516 **16.** 15.99 **17.** 3.68
18. –0.374 **19.** –0.67 **20.** 23.583 **21.** –42.66
22. –23.2 **23.** –0.01 **24.** –2.855 **25.** B **26.** E
27. C **28.** A **29.** A **30.** E **31.** C **32.** E

ANSWERS TO EXERCISES AND PRACTICE TESTS   PREALGEBRA REVIEW

Chapter 3, Section 3 (continued)
33. B **34.** A **35.** D **36.** D **37.** C **38.** B
39. D **40.** A

Section 4
1. 26.23 **2.** 10.92 **3.** 0.35 **4.** 1.08 **5.** 0.519
6. 2.63988 **7.** 33.657225 **8.** 13.736888
9. 0.00048 **10.** 0.054618 **11.** 68.016
12. 88.7315 **13.** 6.56 **14.** 4.31 **15.** 14.045
16. 0.996 **17.** 1.24 **18.** 0.7128 **19.** 0.016
20. 4.25 **21.** 0.045 **22.** 24.26 **23.** 11.8
24. 18.025 **25.** 0.704 **26.** 7.867 **27.** 0.080
28. 8.200 **29.** −2.34 **30.** 5.85 **31.** −4
32. −2.25 **33.** 830 **34.** 0.01256 **35.** 0.305
36. 3.6 **37.** 320 **38.** 0.000144 **39.** −0.000389
40. −1,309,780 **41.** 4.6 **42.** 4.6 **43.** 0.554
44. 20 **45.** 7.44 **46.** 0.056 **47.** D **48.** A
49. E **50.** C **51.** A **52.** B **53.** E **54.** E
55. B **56.** C **57.** A **58.** C **59.** B **60.** D
61. B **62.** B **63.** C **64.** D **65.** E **66.** C

Section 5
1. $\frac{1}{10}$ **2.** $\frac{1}{2}$ **3.** $\frac{3}{4}$ **4.** $\frac{12}{25}$ **5.** $-\frac{3}{5}$
6. $-\frac{7}{10}$ **7.** $-\frac{16}{25}$ **8.** $-\frac{11}{20}$ **9.** $1\frac{1}{5}$ **10.** $4\frac{6}{25}$
11. $\frac{32}{125}$ **12.** $-\frac{19}{40}$ **13.** $-2\frac{2}{125}$ **14.** $7\frac{7}{8}$
15. $\frac{5}{16}$ **16.** $\frac{15}{16}$ **17.** 0.4 **18.** 0.625
19. −0.25 **20.** −0.05 **21.** 3.5 **22.** 2.875
23. 0.36 **24.** 0.90625 **25.** −1.6875 **26.** $0.5\overline{25}$
27. 0.704 **28.** −0.745 **29.** $0.\overline{6}$ **30.** $0.\overline{2}$
31. $−0.1\overline{6}$ **32.** $0.\overline{90}$ **33.** $2.\overline{3}$ **34.** $1.08\overline{3}$
35. $0.8\overline{6}$ **36.** $-0.3\overline{8}$ **37.** $-0.4\overline{09}$ **38.** $0.69\overline{4}$
39. $0.7\overline{1}$ **40.** $0.8\overline{91}$ **41.** $-3.\overline{72}$ **42.** $-5.2\overline{6}$
43. $0.\overline{285714}$ **44.** $0.\overline{692307}$ **45.** 1.44 **46.** 4
47. 3.88 **48.** 0.9 **49.** 0.6 **50.** 8.14 **51.** 0.21
52. 3.15 **53.** 6.34 **55.** −0.3375 **56.** 4.4
57. D **58.** D **59.** B **60.** A **61.** B **62.** A

Section 6
1. 108 **2.** 900 **3.** 1 **4.** 9 **5.** 729 **6.** 16
7. 9^{11} **8.** $(-3)^{17}$ **9.** 7^5 **10.** $(-6)^4$ **11.** 3^{17}
12. $4^1 = 4$ **13.** $\frac{1}{3^4} = \frac{1}{81}$ **14.** $\frac{1}{6^2} = \frac{1}{36}$

15. $\frac{1}{(-2)^5} = -\frac{1}{32}$ **16.** $\frac{1}{(-7)^2} = \frac{1}{49}$ **17.** 9×10^4
18. 6.7×10^8 **19.** 6×10^{-10} **20.** 5.51×10^{-6}
21. -8.893×10^{12} **22.** -4.2×10^{-5}
23. -3.7×10^{-4} **24.** -2.269×10^{-8}
25. 40,000,000 **26.** 75,000,000,000 **27.** 0.00068
28. 0.00000000338 **29.** −1,990,000 **30.** −0.092
31. 6×10^{10} **32.** 6.86×10^{14} **33.** 9.2×10^8
34. 9.76×10^{-3} **35.** 4×10^3 **36.** 1.2×10^8
37. 1.5×10^{-11} **38.** 3.4×10^{18} **39.** 2.8×10^7
40. 6.2×10^{16} **41.** 3×10^3 **42.** 3×10^{-5}
43. 1.15×10^3 **44.** 9.6×10^{-10} **45.** 4.5×10^{-12}
46. 4×10^{13} **47.** 6.25×10^{10} **48.** 1.21×10^{16}
49. 1.1×10^5 **50.** 1.4×10^5 **51.** 1×10^9
52. 1.2×10^{18} **53.** 200,000,000 **54.** 3400
55. 0.0092 **56.** 825 **57.** 19,000 **58.** 0.000013
59. C **60.** D **61.** E **62.** B **63.** A **64.** D
65. A **66.** C **67.** D **68.** B

Section 7
1. 2 **2.** 6 **3.** 11 **4.** 15 **5.** −4 **6.** −10 **7.** $\frac{5}{6}$
8. $\frac{7}{9}$ **9.** $-\frac{1}{5}$ **10.** $-\frac{2}{3}$ **11.** −1.3 **12.** 0.8
13. 0.3 **14.** −1.6 **15.** 20 **16.** −25 **17.** −90
18. 40 **19.** 4 and 5 **20.** 7 and 8 **21.** 10 and 11
22. 9 and 10 **23.** 7 **24.** 8 **25.** 7 **26.** 14
27. −2 **28.** −16 **29.** 15 **30.** 36 **31.** −10
32. −44 **33.** 35 **34.** 3 **35.** −32 **36.** 130
37. 14 **38.** 15 **39.** 20 **40.** 40 **41.** $\sqrt{33}$
42. $\sqrt{105}$ **43.** 3 **44.** 17 **45.** 12 **46.** 21
47. 24 **48.** 13 **49.** 30 **50.** −4 **51.** 14
52. 24 **53.** 4 **54.** 2 **55.** 4 **56.** 7 **57.** $\frac{3}{2}$
58. 2 **59.** 4 **60.** 5 **61.** $\sqrt{7}$ **62.** $\sqrt{7}$
63. $\sqrt{2}$ **64.** $\sqrt{6}$ **65.** D **66.** B **67.** A
68. D **69.** D **70.** C **71.** C **72.** E **73.** B
74. A

PREALGEBRA REVIEW

Chapter 4

Section 1
1. 72   2. –27   3. 0   4. 10   5. –1   6. 29
7. 13   8. 3   9. 1   10. 7   11. $\frac{1}{9}$
12. $\frac{17}{9} = 1\frac{8}{9}$ 13. –8   14. 4   15. 1
16. $-\frac{2}{3}$ 17. –0.4   18. $0.8\overline{3}$, or $\frac{5}{6}$ 19. 48
20. 1   21. $5m, 2n, -3$ 22. $-t^2, \frac{1}{2}t, -4st, 11s, -1$
23. $9x$ 24. $-4n$ 25. $c - 5$ 26. $3a + b$
27. $-6n$ 28. $2x + 2y$ 29. $3a + b$ 30. $-5t + 1$
31. $5xy + x + 3y$ 32. $ab - a$ 33. $6x - 2$
34. $-4x - 20$ 35. $6a + 12b$ 36. $12m - 20n$
37. $-9x - 3$ 38. $10x - 1$ 39. $-17x - 14$
40. $-8y + 30$ 41. $4p - 4q$ 42. $21y - 9x$
43. $5x - 11$ 44. $-p - 7q$ 45. C   46. D   47. A
48. D

Section 2
1. Yes   2. No   3. 6   4. 12   5. 12   6. 0
7. –5   8. –1   9. –2   10. 8   11. –22   12. –10
13. 6   14. 7   15. –9   16. –16   17. $\frac{1}{9}$
18. $-\frac{1}{8}$ 19. $\frac{4}{5}$ 20. $\frac{2}{3}$ 21. 16   22. 24
23. $\frac{3}{10}$ 24. $-\frac{9}{7} = -1\frac{2}{7}$ 25. –25   26. 28
27. $-\frac{21}{8} = -2\frac{5}{8}$ 28. $-\frac{1}{8}$ 29. $\frac{5}{4} = 1\frac{1}{4}$
30. $-\frac{17}{5} = -3\frac{2}{5}$ 31. $\frac{3}{2} = 1\frac{1}{2}$ 32. $-\frac{35}{24} = -1\frac{11}{24}$
33. 2.1   34. 4.2   35. 5   36. –3.15   37. 1.2
38. 1.25

Section 3
1. 4   2. 1   3. –4   4. –2   5. $\frac{3}{2} = 1\frac{1}{2}$
6. $-\frac{5}{3} = -1\frac{2}{3}$ 7. 8   8. –1   9. –4   10. 8
11. –30   12. –56   13. $\frac{3}{7}$ 14. $\frac{3}{32}$
15. $\frac{3}{2} = 1\frac{1}{2}$ 16. $\frac{7}{2} = 3\frac{1}{2}$ 17. 0.4   18. 0.3

ANSWERS TO EXERCISES AND PRACTICE TESTS

19. –4.5   20. –9.8

Section 4
1. $9 - n$ 2. $12n$ 3. $n + 20$ or $20 + n$ 4. $4 + n$ or $n + 4$ 5. $\frac{16}{n}$ 6. $\frac{3}{4}n$ 7. $n - 1$ 8. $\frac{n}{8}$
9. $5n - 2$ 10. $6 + \frac{n}{3}$ 11. $\frac{1}{2}(n + 13)$
12. $4(7 - n)$ 13. $x - 14 = 7$ 14. $18 = \frac{2}{3}x$
15. $3x - 5 = 10$ 16. $\frac{x}{8} + 1 = 9$
17. $16 = 2(x + 12)$ 18. $\frac{(x-6)}{30} = 8$
19. $9x - 5 = 14$ 20. $-4x + 25 = 2$ 21. C   22. B
23. E   24. B   25. A   26. B

Chapter 5

Section 1
1. 2 to 3, 2:3, $\frac{2}{3}$ 2. 15 to 30, 15:30, $\frac{15}{30} = \frac{1}{2}$
3. 7 to 10, 7:10, $\frac{7}{10}$ 4. 1 to 5, 1:5, $\frac{1}{5}$ 5. $\frac{3}{4}$
6. $\frac{4}{3}$ 7. $\frac{3}{5}$ 8. $\frac{1}{4}$ 9. $\frac{1}{2}$ 10. $\frac{31}{11}$ 11. $\frac{17}{32}$
12. $\frac{3}{2}$ 13. $\frac{4}{7}$ 14. $\frac{5}{8}$ 15. $\frac{4}{5}$ 16. $\frac{10}{3}$
17. B   18. A   19. B   20. D   21. C   22. C
23. E   24. A   25. C   26. A

Section 2
1. $\frac{32 \text{ meters}}{1 \text{ second}}$ 2. $\frac{12 \text{ cans}}{1 \text{ box}}$ 3. $\frac{100 \text{ miles}}{3 \text{ hours}}$
4. $\frac{6 \text{ ounces}}{1 \text{ serving}}$ 5. $\frac{\$5}{4 \text{ dozen}}$ 6. $\frac{1 \text{ bus}}{46 \text{ people}}$
7. $\frac{3 \text{ instructors}}{55 \text{ students}}$ 8. $\frac{\$41}{2 \text{ hours}}$ 9. $\frac{3 \text{ cookies}}{1 \text{ child}}$
10. $\frac{33 \text{ miles}}{1 \text{ gallon}}$ 11. 25 mg per pill
12. 2304 people per sq mi   13. 4 hrs per day
14. 20 min per period   15. 2.5 yds per down
16. $80\frac{1}{4}$ bushels per acre   17. $3.49 per lb

Chapter 5, Section 2 (continued)
18. $2.15 per gal   19. 5.5 yds per dress
20. $94 per person   21. a 12 oz cup
22. 20 lb bag   23. 14 oz can   24. 120 sheets
25. Ted   26. mid-size car   27. C   28. B   29. C
30. E   31. D   32. B   33. C   34. B   35. B
36. D

Section 3
1. $\frac{3}{4} = \frac{15}{20}$ 2. $\frac{18}{32} = \frac{9}{16}$ 3. $\frac{8}{24} = \frac{20}{60}$
4. $\frac{21}{42} = \frac{7}{14}$ 5. $\frac{4\frac{1}{2}}{9} = \frac{1\frac{1}{8}}{2\frac{1}{4}}$ 6. $\frac{6.25}{11} = \frac{3.125}{5.5}$
7. $\frac{\$165}{5 \text{ hours}} = \frac{\$105}{3 \text{ hours}}$ 8. $\frac{208 \text{ miles}}{8 \text{ gallons}} = \frac{286 \text{ miles}}{11 \text{ gallons}}$
9. $\frac{32 \text{ feet}}{1 \text{ second}} = \frac{288 \text{ feet}}{9 \text{ seconds}}$
10. $\frac{25 \text{ yards}}{75 \text{ feet}} = \frac{16 \text{ yards}}{48 \text{ feet}}$ 11. True   12. True
13. Not true   14. True   15. True   16. True
17. Not true   18. Not true   19. True   20. True
21. True   22. Not true   23. True   24. True
25. Not true   26. True

Section 4
1. 3   2. 18   3. 49   4. 5   5. 4   6. 9   7. 12
8. 18   9. $\frac{54}{5} = 10\frac{4}{5}$ 10. $\frac{16}{3} = 5\frac{1}{3}$
11. $\frac{63}{10} = 6\frac{3}{10}$ 12. $\frac{3}{2} = 1\frac{1}{2}$ 13. 15   14. 28
15. 15   16. 20   17. 8   18. 150   19. B   20. E
21. B   22. A   23. B   24. C   25. C   26. C
27. D   28. B

Section 5
1. B   2. C   3. D   4. A   5. B   6. E   7. C
8. D   9. B   10. B   11. E   12. C   13. A
14. B   15. E   16. B   17. C   18. B   19. C
20. A   21. C   22. D   23. E   24. B   25. A
26. C   27. D   28. B

Chapter 6

Section 1
1. 68%   2. 1%   3. 83%   4. 15%   5. 91%
6. 43%   7. $\frac{23}{100}$ 8. $\frac{61}{100}$ 9. $\frac{18}{100}$ 10. $\frac{57}{100}$
11. $\frac{6}{100}$ 12. $\frac{89}{100}$ 13. 74%   14. 32%
15. 12%   16. 81%   17. Brand B   18. Computer City   19. Midstate Bank   20. topical ointment Y

Section 2
1. 0.72   2. 0.13   3. 0.9   4. 0.1   5. 0.05
6. 0.03   7. 1.5   8. 1.33   9. 0.048   10. 0.0967
11. 0.002   12. 0.0004   13. 0.1835   14. 0.681
15. 0.00016   16. .00207   17. 48%   18. 22%
19. 70%   20. 60%   21. 9%   22. 1%
23. 37.5%   24. 90.3%   25. 166%   26. 210%
27. 3.3%   28. 0.9%   29. 55.15%   30. 7.18%
31. 0.46%   32. 0.08%   33. C   34. A   35. D
36. B

Section 3
1. $\frac{9}{25}$ 2. $\frac{19}{20}$ 3. $\frac{1}{5}$ 4. $\frac{3}{10}$ 5. $\frac{9}{4}$ 6. $\frac{11}{10}$
7. $\frac{3}{50}$ 8. $\frac{1}{20}$ 9. $\frac{16}{125}$ 10. $\frac{261}{500}$ 11. $\frac{7}{8}$
12. $\frac{1}{400}$ 13. $\frac{4}{9}$ 14. $\frac{2}{3}$ 15. $\frac{63}{125}$ 16. $\frac{1}{200}$
17. 50%   18. 75%   19. 62.5%   20. 65%
21. 56.25%   22. 66%   23. 70%   24. 96%
25. 150%   26. 137.5%   27. 210%   28. 180%
29. 16.67%   30. 8.33%   31. 88.89%
32. 86.67%   33. 57.14%   34. 81.82%
35. 62.22%   36. 38.89%

Section 4
1. 56%   2. 54   3. 18   4. 75   5. 68%   6. 200
7. 150%   8. 101.4   9. 300%   10. 1%   11. 48
12. 339.6   13. B   14. A   15. E   16. B   17. C
18. D   19. A   20. D

PREALGEBRA REVIEW ANSWERS TO EXERCISES AND PRACTICE TESTS

Section 5
1. B 2. D 3. A 4. B 5. C 6. D 7. E
8. A 9. C 10. E 11. C 12. A 13. D
14. A 15. B 16. C 17. B 18. D 19. E
20. C 21. B 22. C 23. D 24. B 25. C
26. B 27. E 28. C 29. B 30. B 31. A
32. C 33. E 34. A 35. D 36. C 37. D
38. A 39. B 40. C 41. C 42. D

Section 6
1. 124 2. 0.43 3. 86 4. 75° 5. 15
6. $71.29 7. 16 8. 1924 9. 40 10. 98
11. 5 hrs 12. 3 in 13. 132 14. 5.9 15. no mode 16. 89 and 96 17. 8 18. 156

Answers to Practice Tests

Practice Test A
1. B 2. B 3. D 4. C 5. A 6. B 7. E
8. A 9. E 10. C 11. B 12. D 13. A
14. E 15. C

Practice Test B
1. A 2. C 3. B 4. A 5. D 6. C 7. B
8. C 9. B 10. C 11. A 12. B 13. C
14. D 15. B

Practice Test C
1. C 2. A 3. E 4. C 5. B 6. A 7. D
8. D 9. B 10. E 11. C 12. A 13. E
14. C 15. D